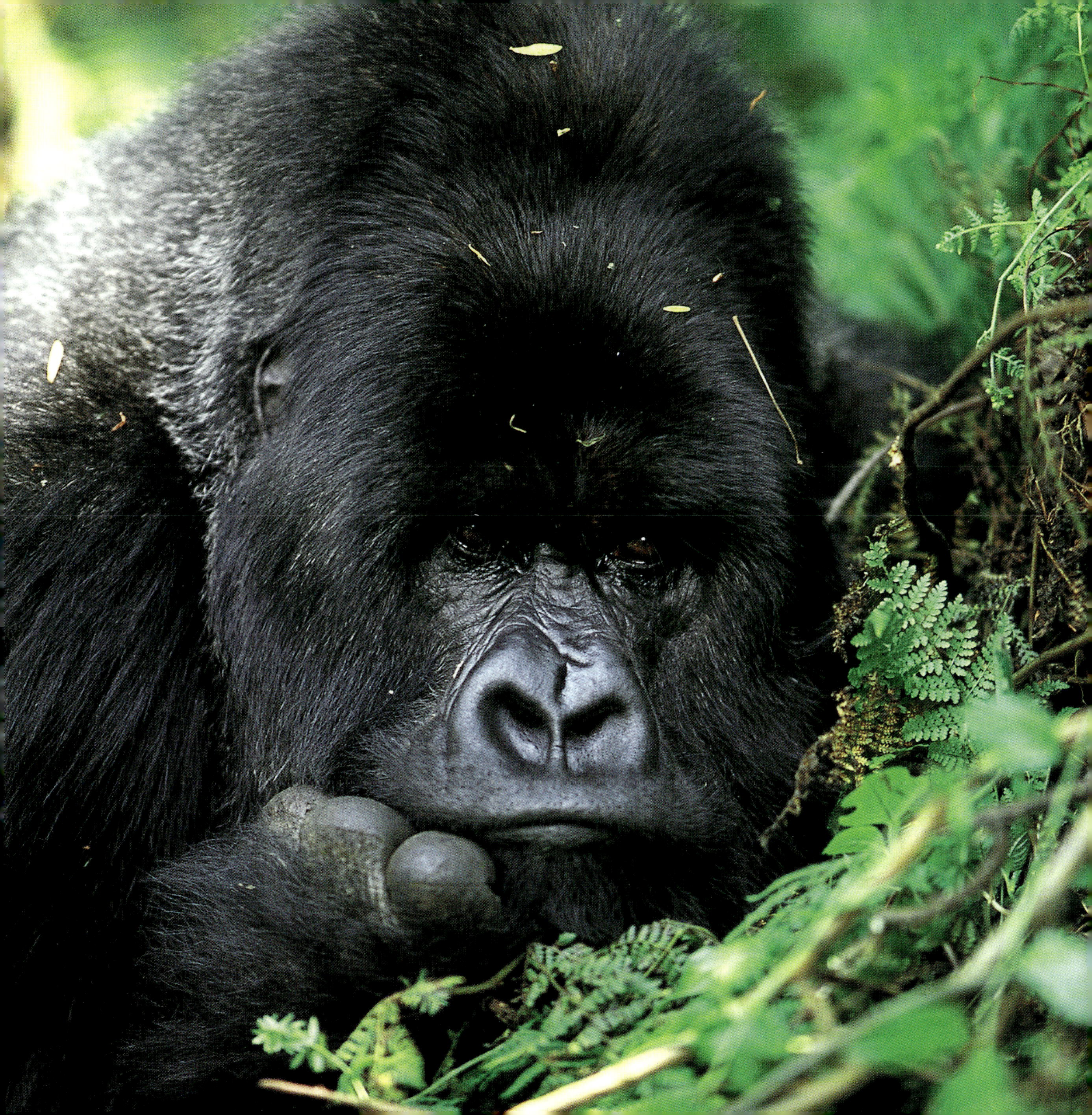

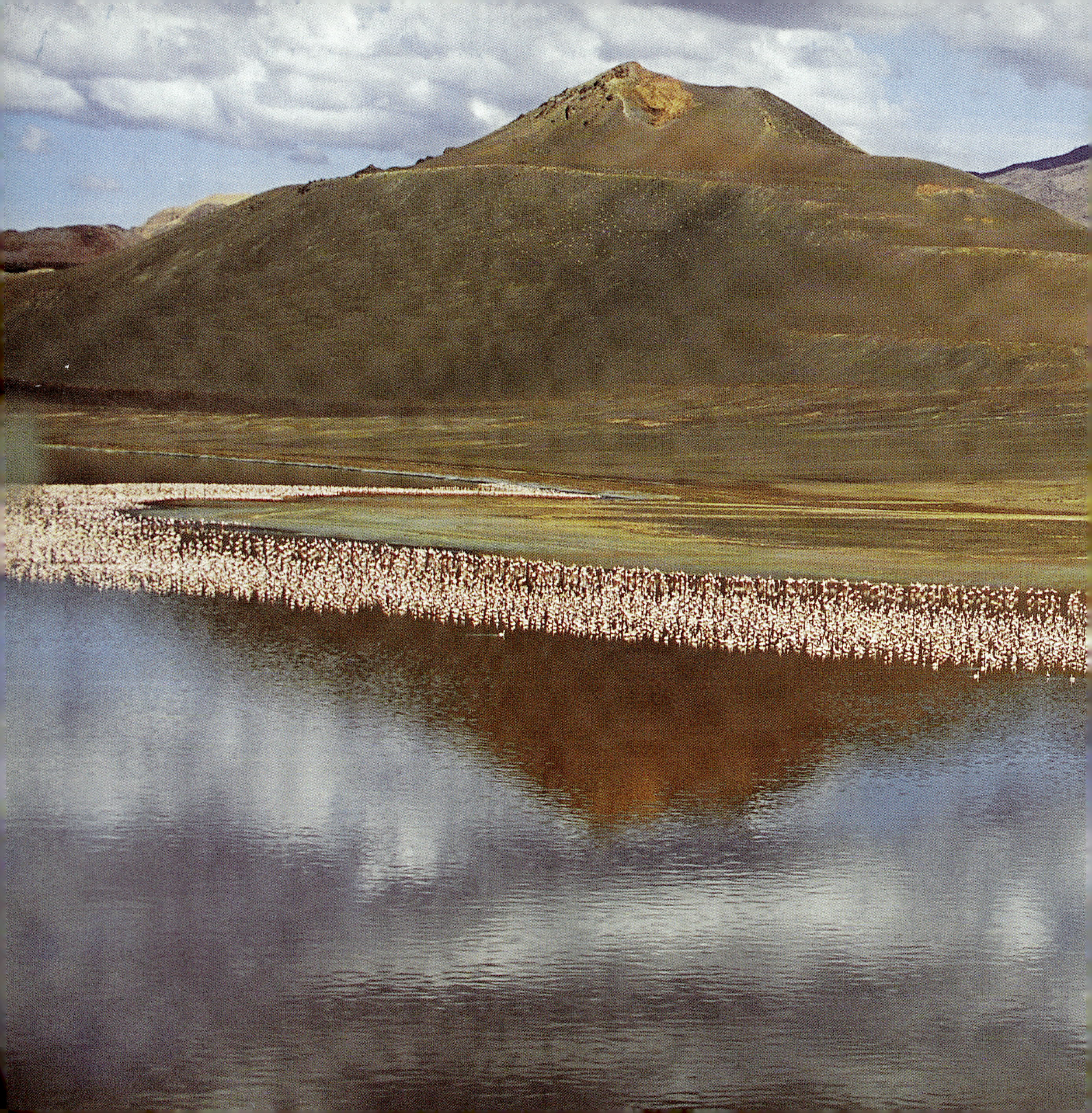

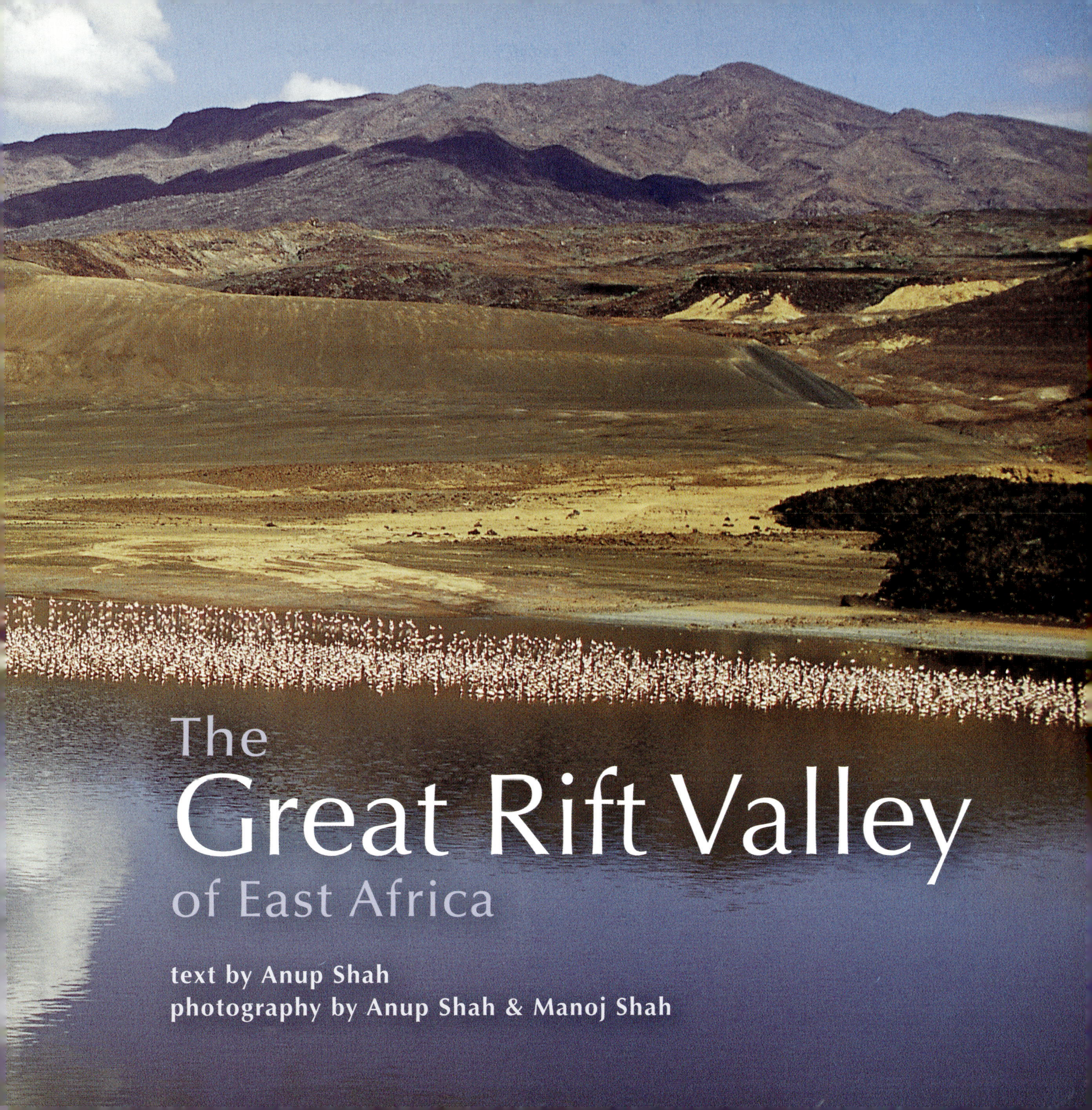

The Great Rift Valley of East Africa

text by Anup Shah
photography by Anup Shah & Manoj Shah

Contents

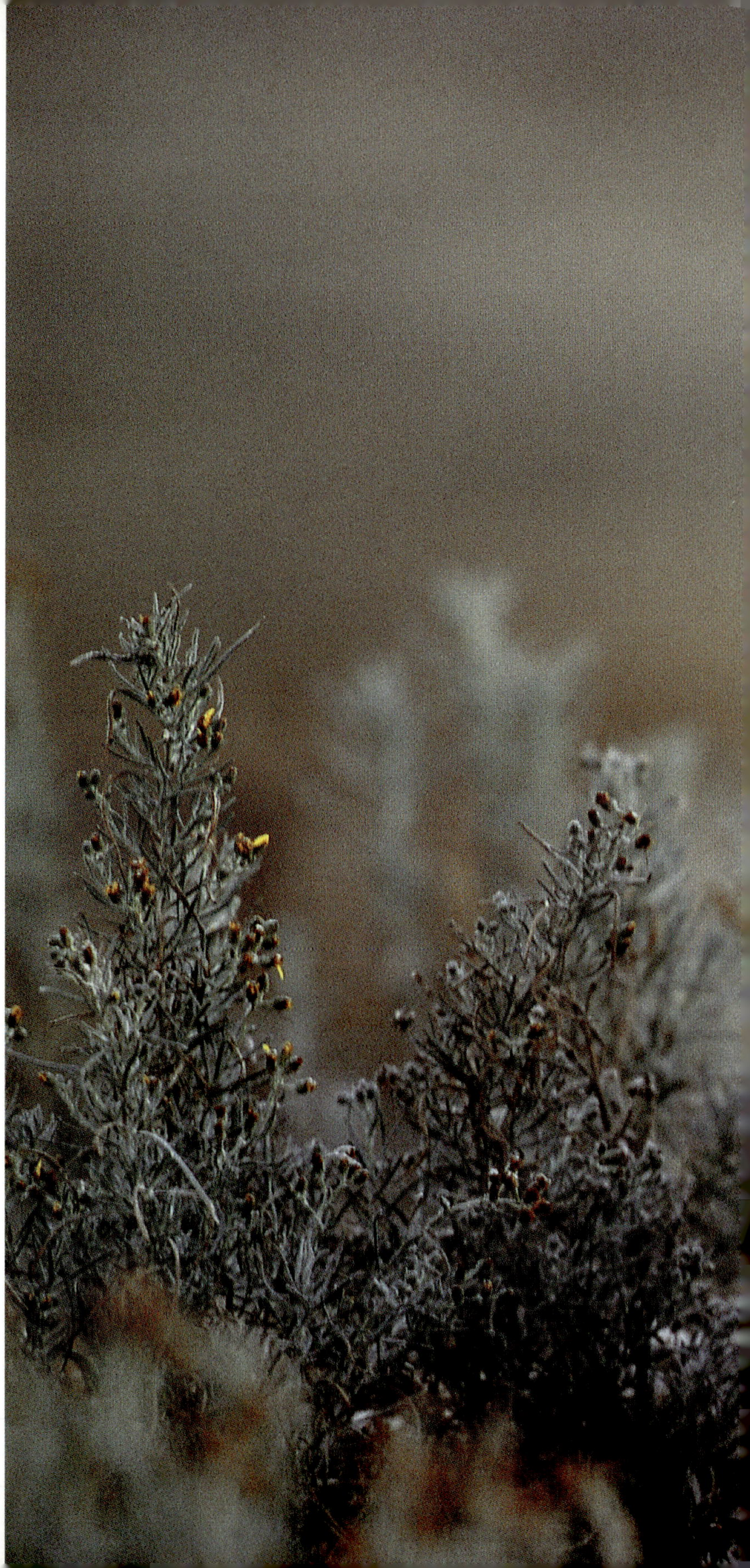

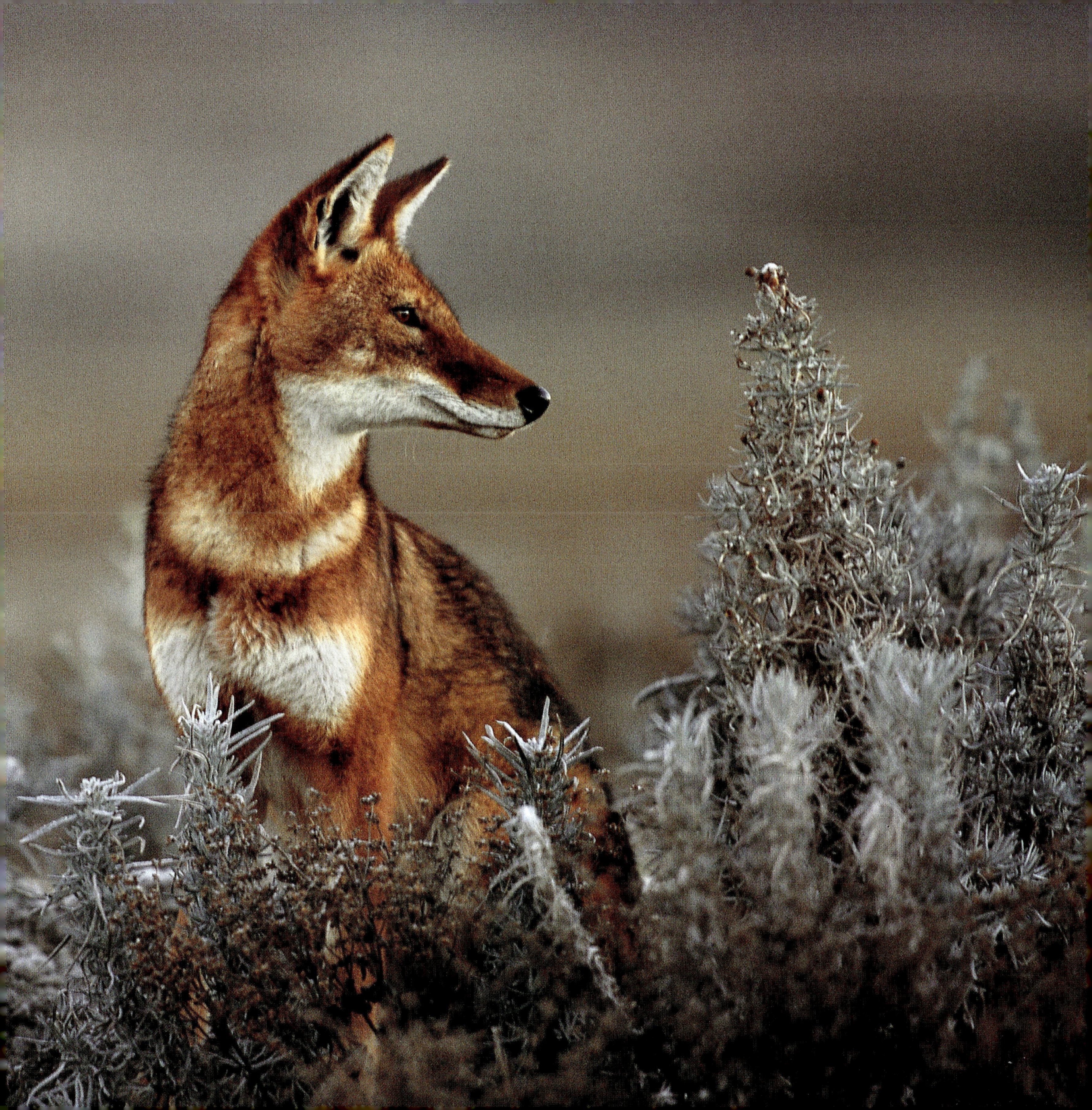

DEDICATION

To the memory of our mother,
Zaverben Shah, who taught us virtue.

Struik Publishers
(a division of New Holland Publishing (South Africa) (Pty) Ltd)
Cornelis Struik House
80 McKenzie Street
Cape Town 8001

New Holland Publishing is a division of Avusa Ltd

Visit us at **www.struik.co.za**
Log on to our photographic website **www.imagesofafrica.co.za**
for an African experience

First published in 2008

Publishing manager: Pippa Parker
Managing editor: Helen de Villiers
Editor: Leonie Hofmeyr-Juritz
Consultant: Gordon Boy
Designers: Janice Evans and Robin Cox
Cartographer: Martin Endemann
Indexer: Cora Ovens
Proofreader: Glynne Newlands

Reproduction by Hirt & Carter Cape (Pty) Ltd
Printed and bound by Craft Print International Ltd

ISBN 978 1 77007 450 7

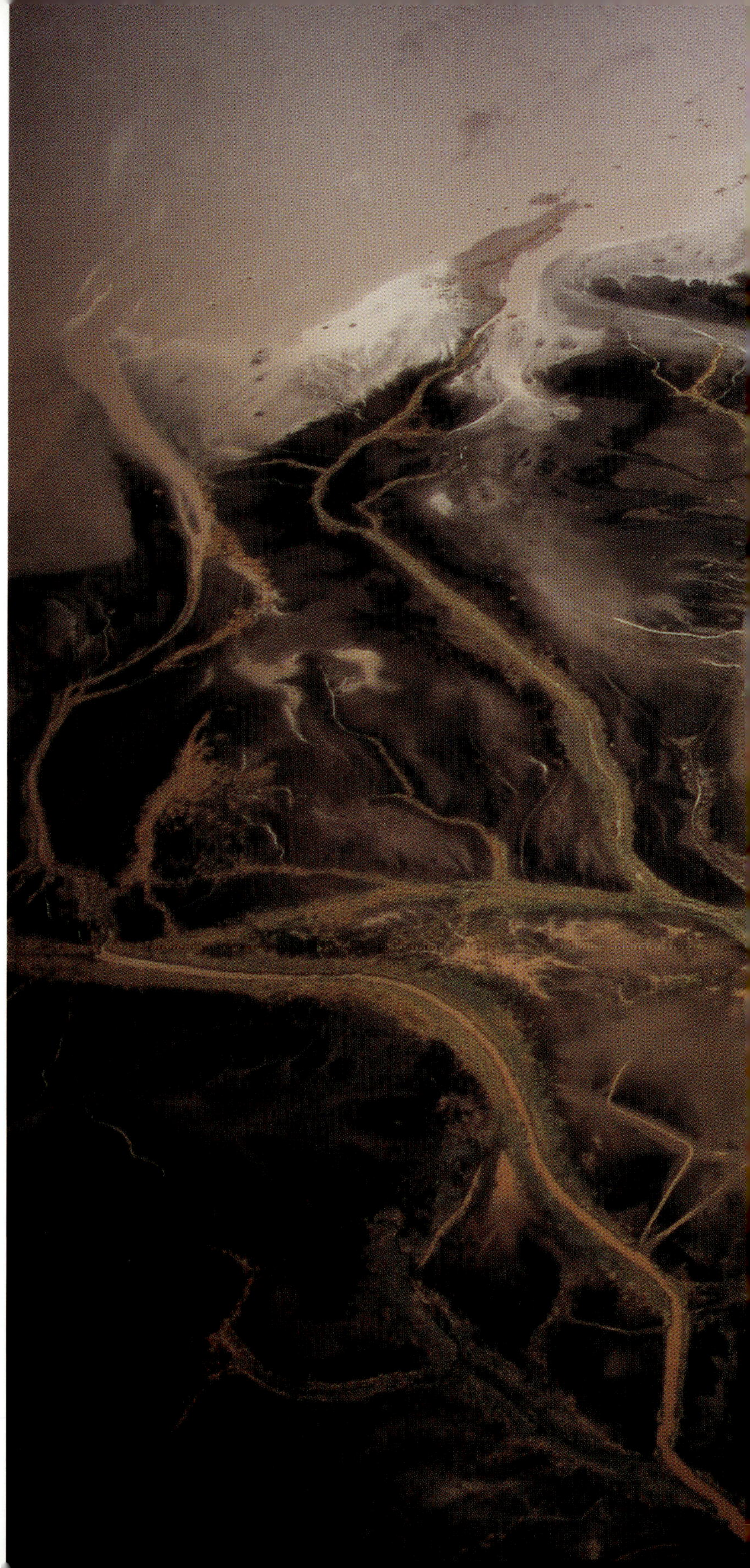

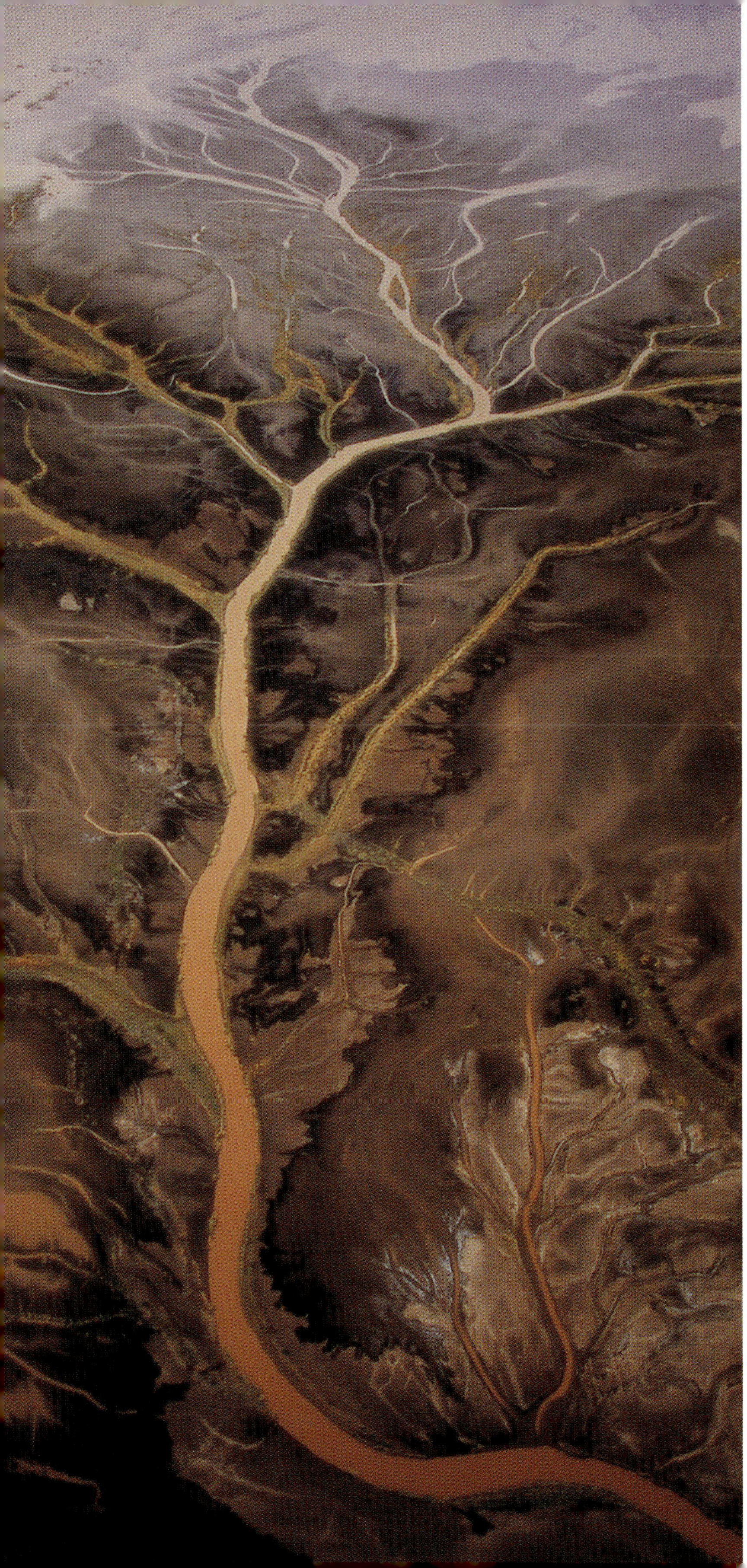

Foreword

When we were little boys in shorts, our parents would drive us 40 kilometres from Nairobi, right up to the eastern wall of the Great Rift Valley, and stop at a place popularly known as 'The Escarpment'. Here, we would noisily tumble out of the car and then gaze, spellbound, at the valley floor below, the forbidding Longonot Volcano, and Lake Naivasha shimmering in the distant heat. On a clear day, we could also make out the western wall of the rift valley and Suswa Volcano. The mysterious and hypnotic scene, reminiscent of the Pleistocene era, inspired in us flights of vivid imagination. What secrets were locked in this seemingly lost world?

As children with more imagination than knowledge, we gravely underestimated the treasures locked in the great valley. Armed only with local knowledge, we had no conception of the size, scale and scope of this great fissure in the Earth. We failed to appreciate the diversity of habitats and the variety of life forms contained in these vast landscapes. Little did we realise that this wild and pristine place was the site of our human origins. Much wiser now, we hope this book does belated justice to the awe-inspiring wonder of the Great Rift Valley.

Page 1: A Virunga silverback mountain gorilla, Gorilla gorilla beringei*, seemingly lost in thought.*

Title pages: Lake Logipi, lying on the rift valley floor in northern Kenya, is a seasonal lake.

Previous spread: An Ethiopian wolf in Bale Mountains National Park surveys its domain.

Left: A freshwater river flows on the rift valley floor. The 'bird's foot' delta is formed by an accumulation of debris and salt.

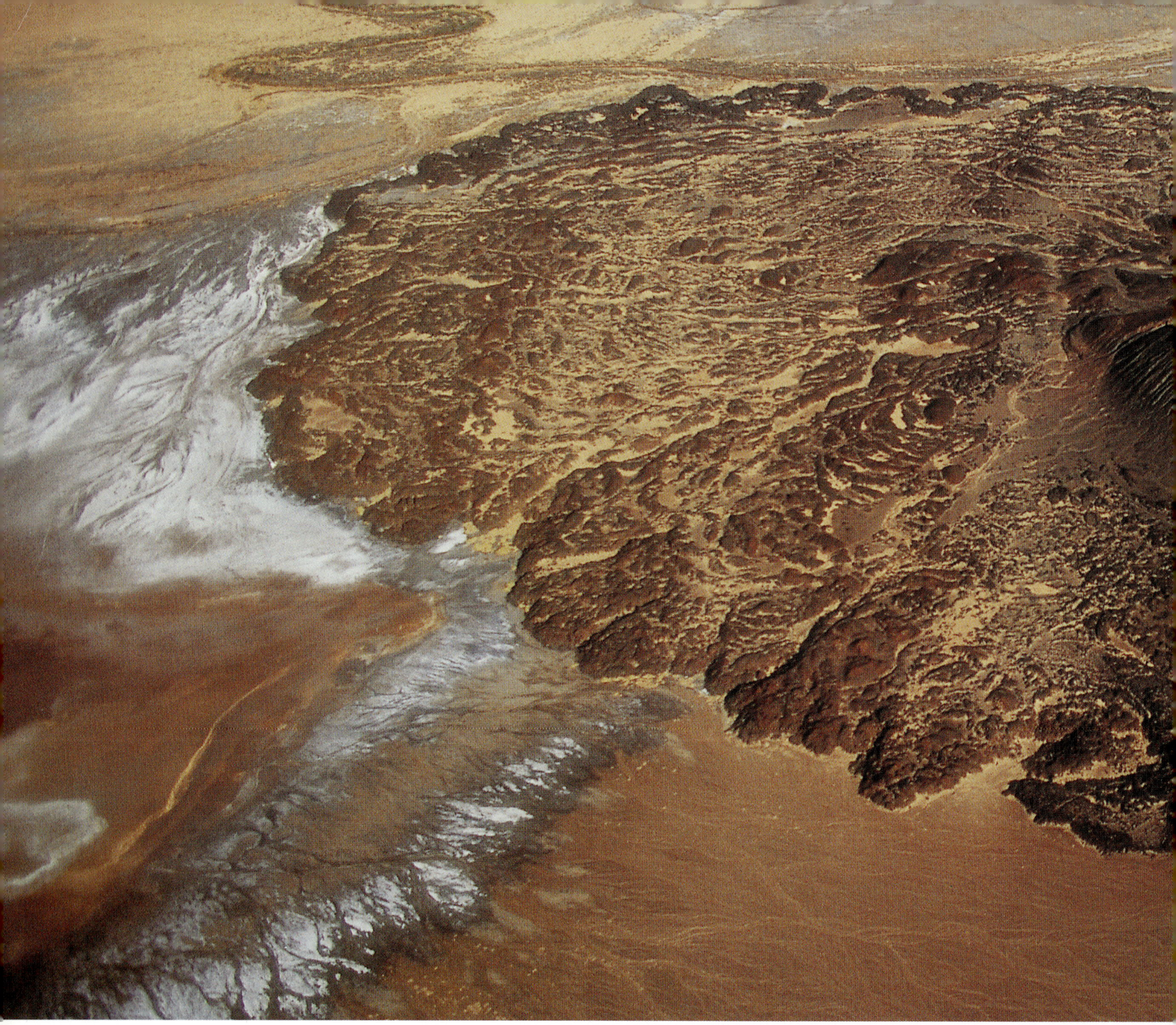

An aerial view of an extinct volcano on the rift valley floor, formed during the dying, quiescent phase of the Kenyan section of the valley.

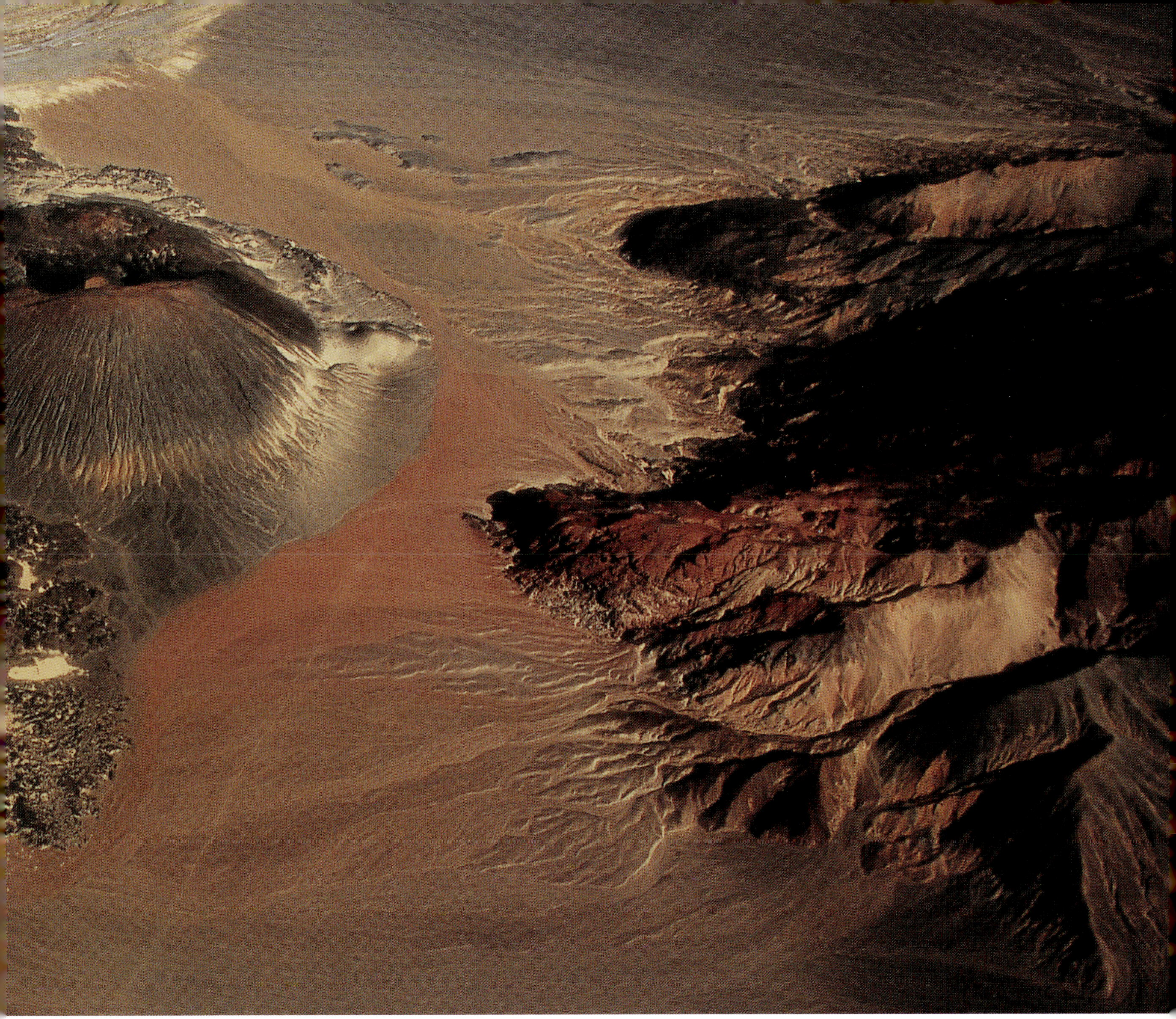

Introduction

Nabuyatom Volcano, at the edge of Lake Turkana in northern Kenya, is an almost uneroded caldera, more than one kilometre wide. Black lava covers the land as far as Teleki's Volcano further south.

When Neil Armstrong landed on the moon and looked back at the spherical Earth, the terrestrial feature that he most easily recognised was the Great Rift Valley of Africa. A relatively new continental rift, this valley stretches in a sequence of subsidences for some 6 400 kilometres (4 000 miles), from Lebanon in the north to Mozambique in the south. The rifting process began about 40 million years ago when the crust and lithosphere of the tectonic plate here were thinned and stretched by powerful convection currents in the underlying mantle, which rose up to form an elevated tract. Major faulting in the crust's rocks , thought to have occurred 11 million years ago, produced a rift or fault line. As the areas to either side of the fault line moved away, the land between its walls dropped down to form a vast valley, characterised by numerous lake basins, volcanoes and escarpments.

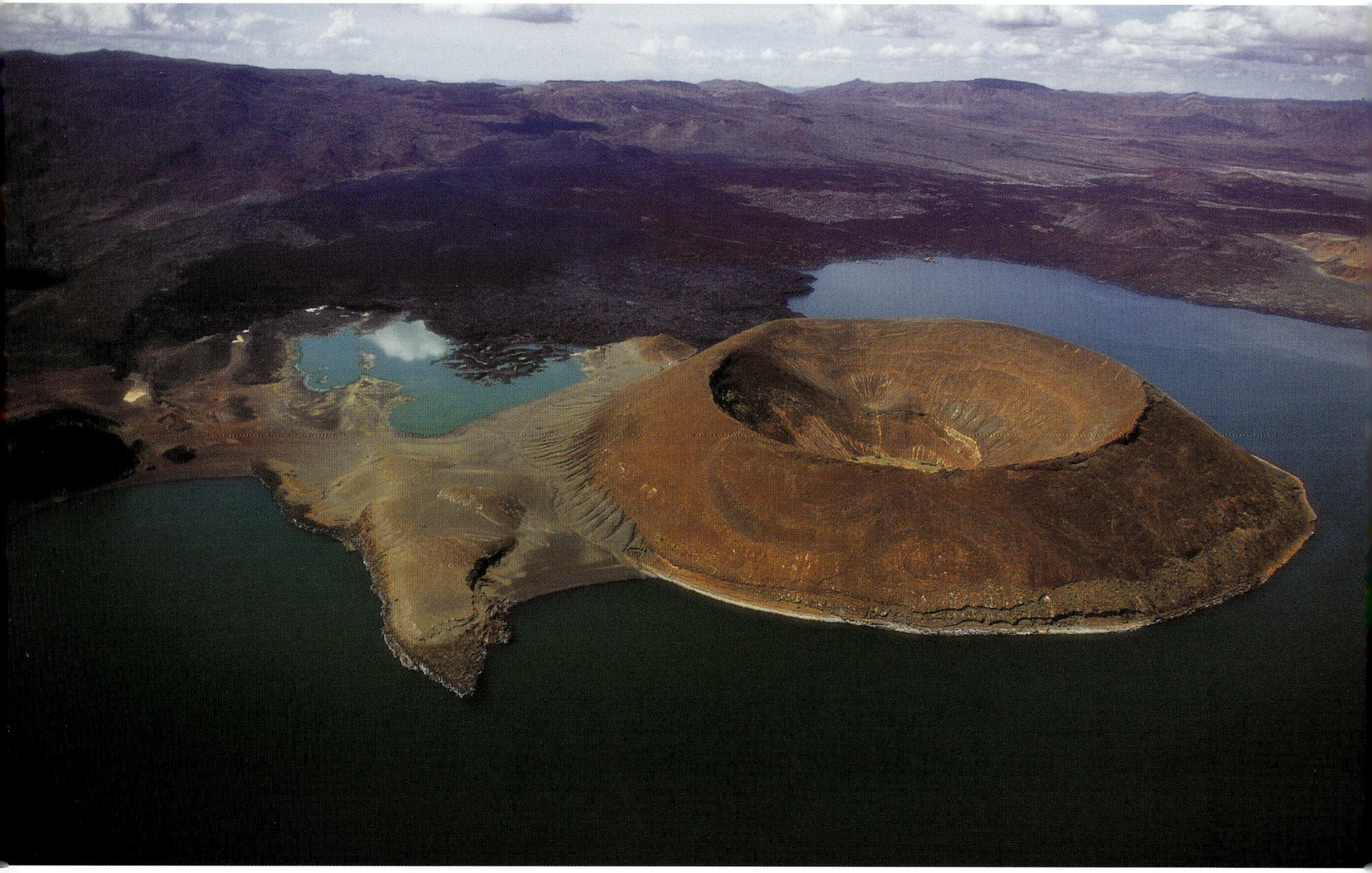

Rift valley geography

From the Jordan Valley, the rift valley runs through the ocean-filled trench of the Red Sea, reaching Africa in the low-lying Danakil Desert of northern Ethiopia. From here it continues as a huge chasm, only rarely wider than 48 kilometres (30 miles), climbing and plunging through the landscape of East Africa. It rises steeply into the Ethiopian Highlands, and continues south at 1 860 metres (6 000 feet) above sea level, its course marked by a series of lakes where the walls of the valley are broken by cross-fractures. In Kenya, it proceeds through Lake Turkana, rising further to Lake Naivasha – the highest lake in the valley – and beyond, to a further series of lakes, before descending into Tanzania and Mozambique. Heading toward the Indian Ocean, it eventually culminates beyond Lake Malawi.

From this great fissure in the Earth's crust, there are other lesser cracks that branch off: the Gulf of Suez and the Gulf of Aqaba stretch out from the northern end of the Red Sea, and the Gulf of Aden from its southern end; in Africa, a lesser western rift containing, among other lakes, Lake Tanganyika, curves northward from Lake Malawi.

A striking feature of the Great Rift Valley is that, within its confines, there are innumerable smaller fault scarps where the land has moved up and down. These form a complex landscape of cliffs and lake basins, dotted with volcanoes, within which a great diversity of species has evolved and continues to flourish.

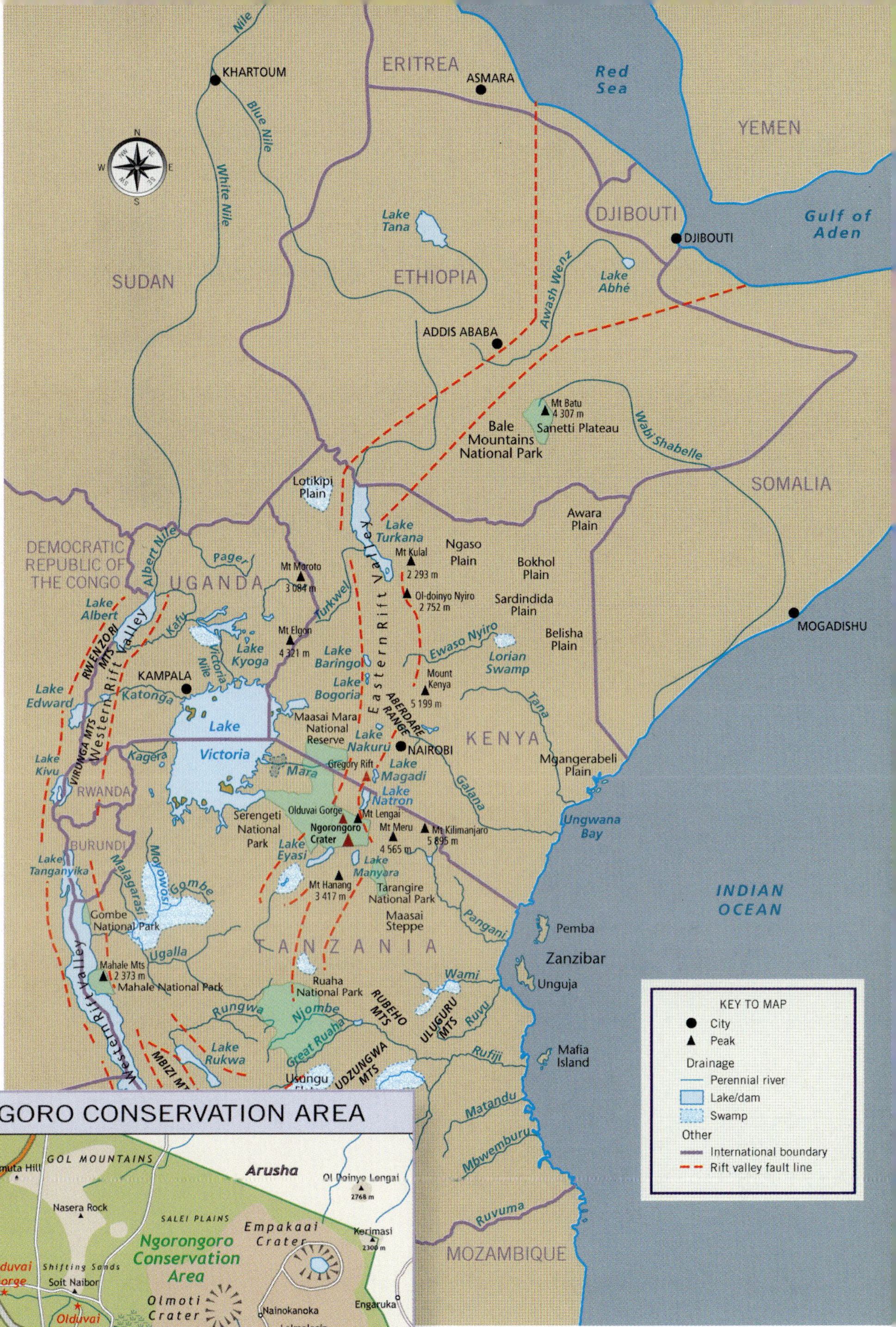

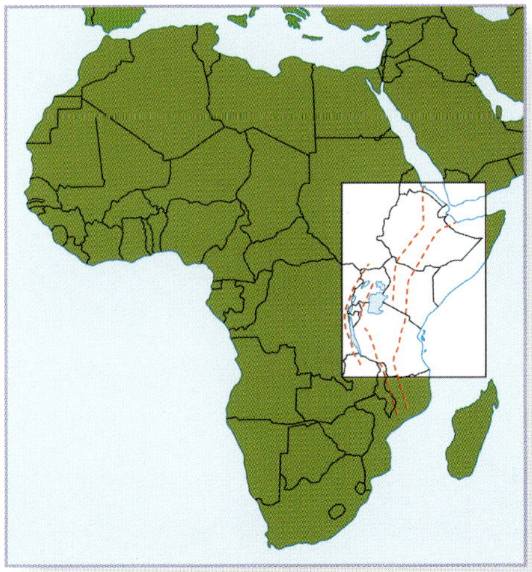

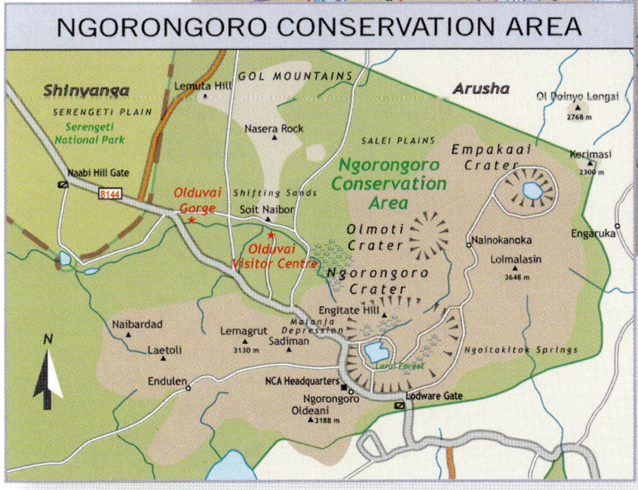

Above: The eastern section of the Great Rift Valley.

Left: Detail of the Ngorongoro Conservation Area.

Far left: The Great Rift Valley.

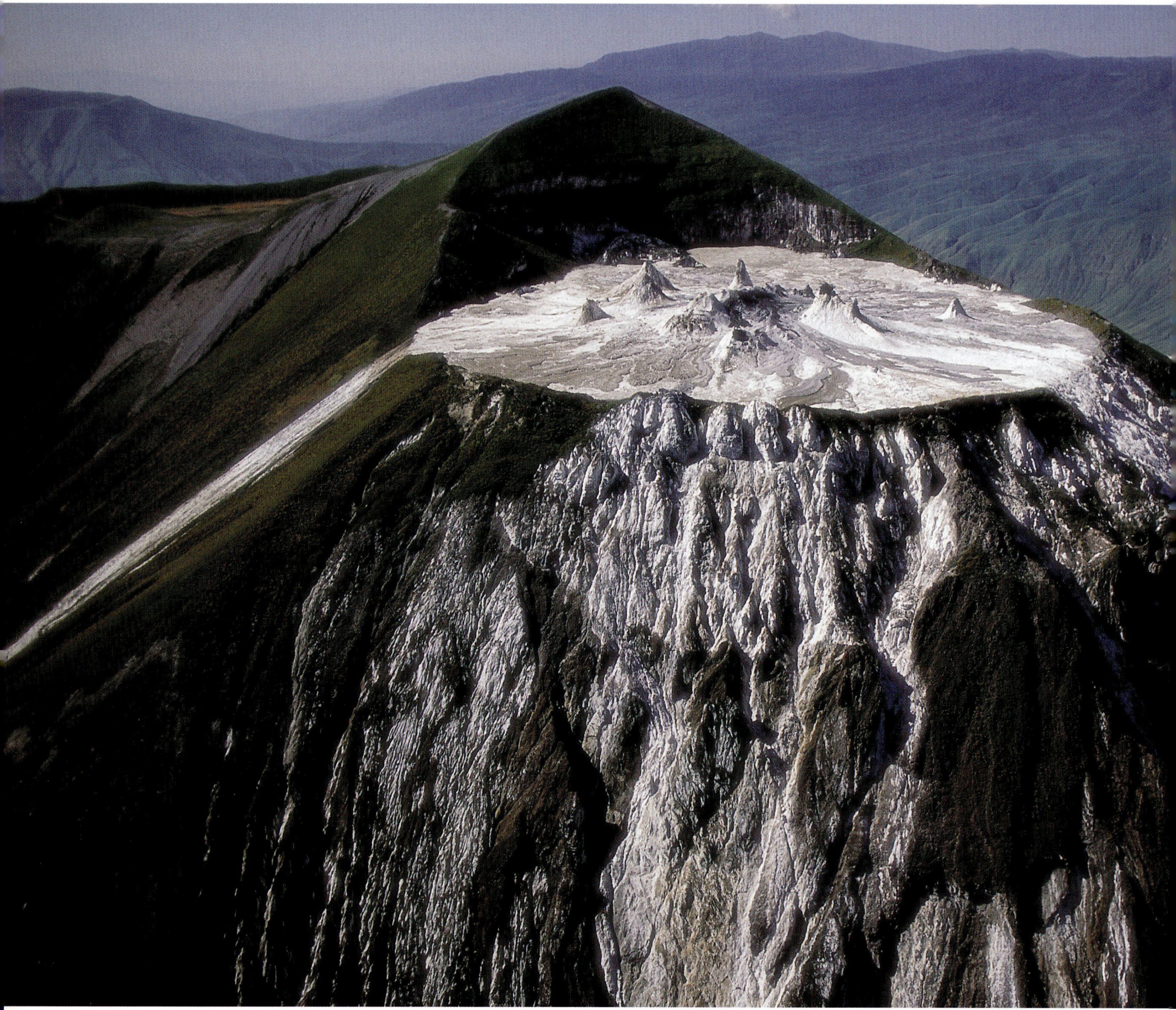

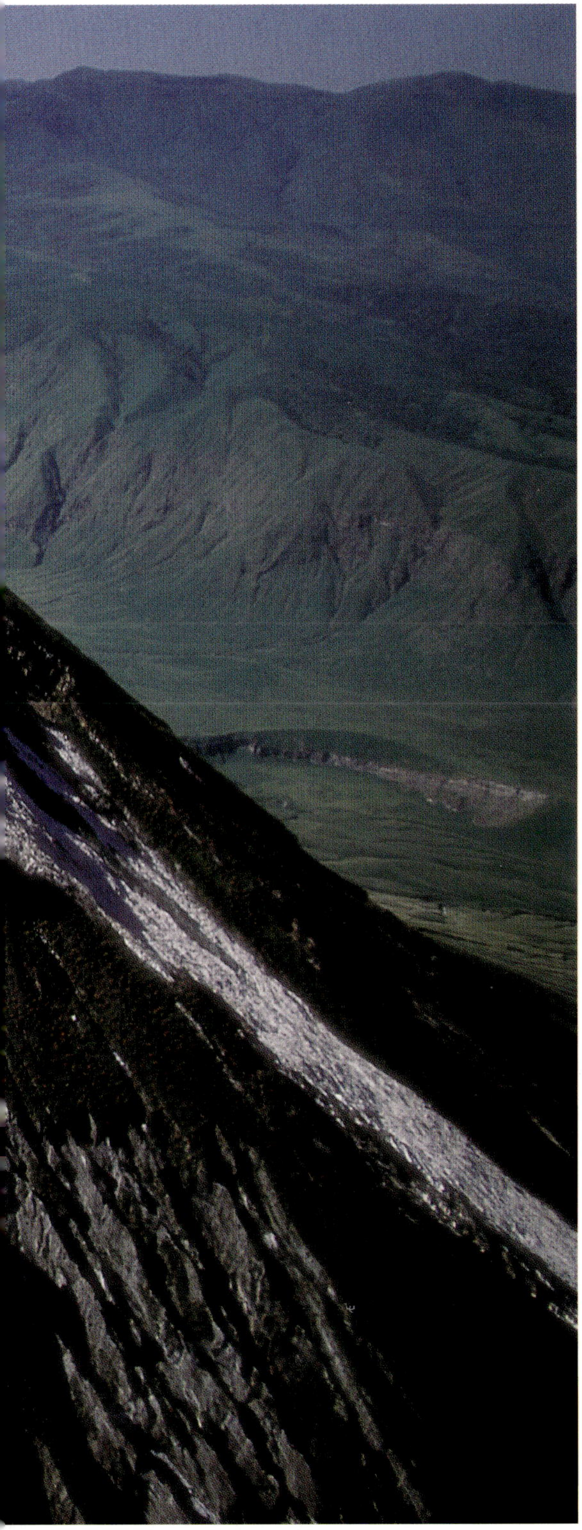

Evolution of life

For the first few billion years after the Earth was formed from the sun, no life existed. It is believed that life began, three to four billion years ago, from a self-replicating organic molecule distantly related to the four molecules that make up the genetic code. Although we do not know exactly how and why life began, there is much evidence to support its subsequent evolution.

Monkeys and apes

Fast tracking through time, several billion years of evolution passed; the dinosaurs came and went, and the mammals started to proliferate. Among these, the monkeys and apes lived in forest habitats, where they used their forelegs to swing through the forest canopy – but they sometimes came down from the trees to forage for food on the forest floor. When they walked they used their forelegs the way other animals do; but when climbing trees, picking fruit or catching small animals on the ground, they used the extremities of their forelimbs in specialised ways – as hands.

Left: *With its unmistakable cone shape, Tanzania's volcanic mountain, Ol Doinyo Lengai, has steep and deeply eroded sides. When Lengai erupts, it pours out black lava that is chemically unstable. Within 48 hours of coating the slopes, the lava turns white as contact with moisture in the air converts much of it to crystals of sodium carbonate, or 'washing soda'.*

Below: *Mountain gorillas,* Gorilla gorilla beringei*, live on the slopes of the volcanic Virunga Mountains, situated in the western arm of the rift valley, straddling Uganda, Rwanda and the DR Congo.*

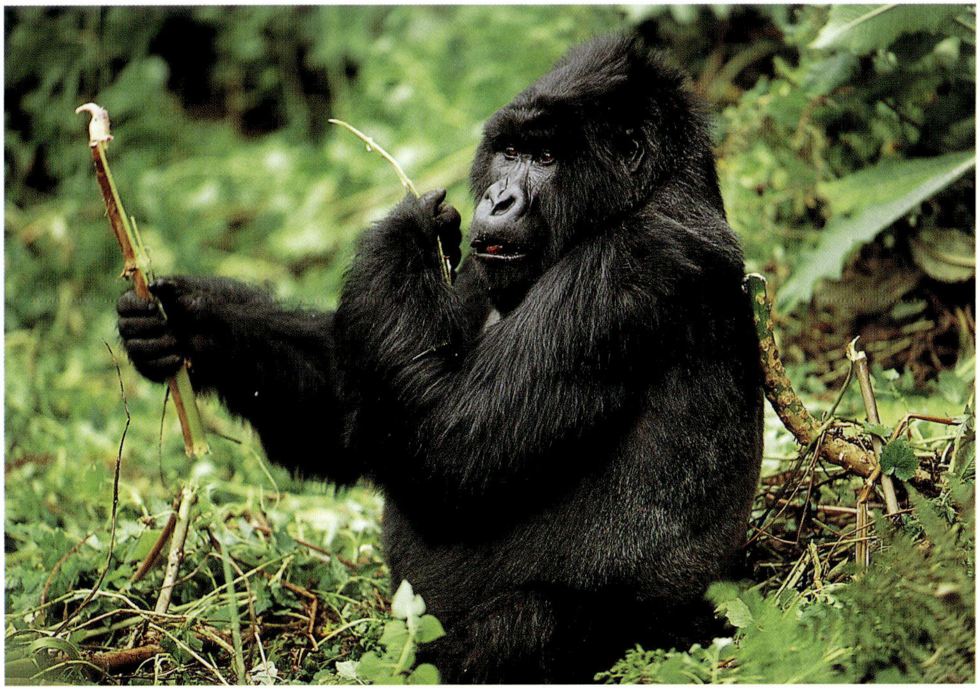

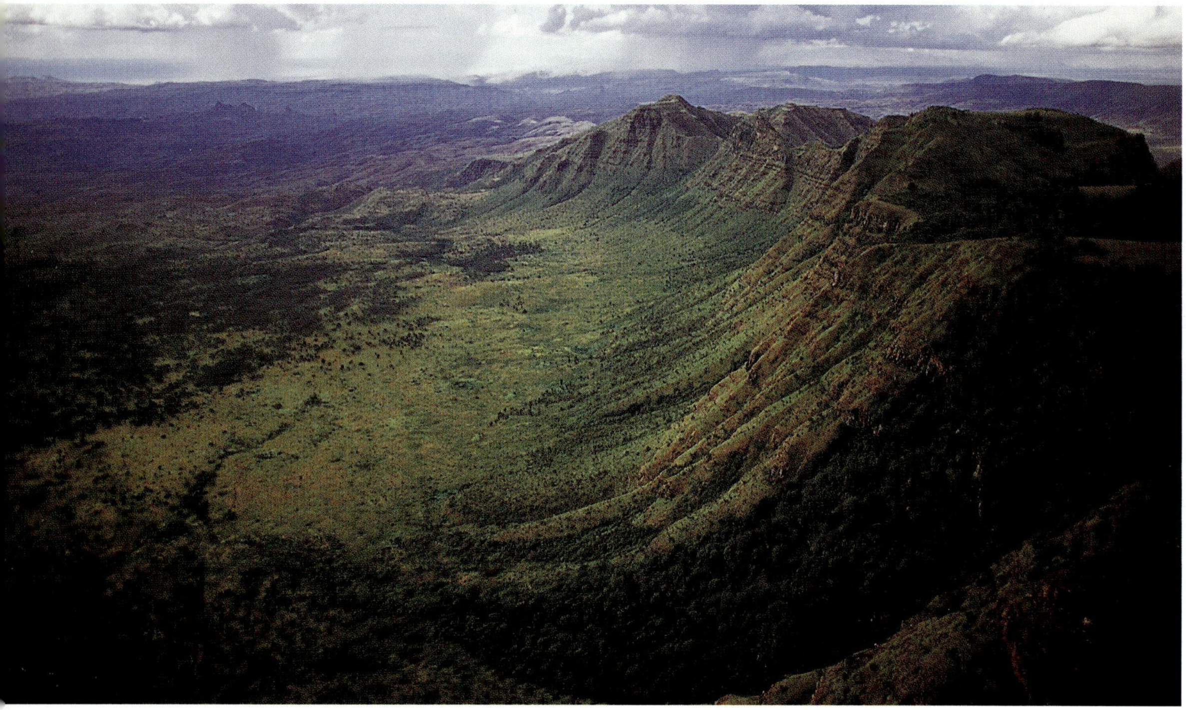

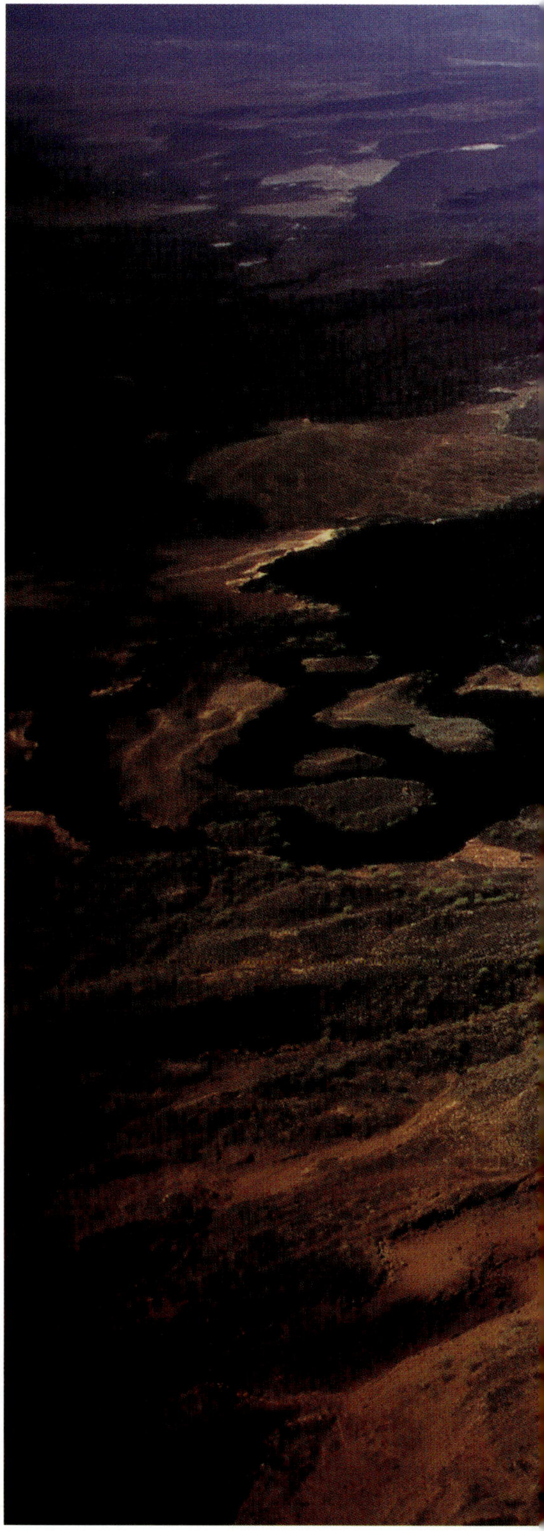

Above: *The eastern wall of the Great Rift Valley.*
Right: *In 1935, Lieutenant Andrew saw a smouldering cone on the southern slope of a volcanic ridge and named it Andrew's Volcano. It is now dormant; a few steam jets are the only sign of latent activity.*

Some of the early apes and monkeys were already in place when the Great Rift Valley began to take shape. A world of new landscapes was created, as existing environments were quite literally fragmented to create a diversity of 'evolutionary hotspots' – places where evolutionary events tend to occur relatively frequently. The dynamic tectonic changes that characterised the formation of the Great Rift Valley accelerated evolutionary change and speciation. Nowhere is this more spectacularly demonstrated than in the eastern African sections of the valley – from the Ethiopian Highlands in the north to Tanzania's Crater Highlands, and from Lake Albert on the Uganda-DR Congo border in the west to Lake Tanganyika in the south. This region sustains an extraordinarily rich fauna – including the Ethiopian wolf that is endemic to the northern highlands, the flamingos of the burning soda lakes of Kenya and Tanzania, the huge herbivore populations of northern Tanzania's ash-covered plains, and – holed up in the fragmented forests of the western rift – remnant populations of two of the world's great apes, the mountain gorilla and the chimpanzee. The Great Rift Valley's diverse, interacting habitats have done much to shape the evolution and survival of such life forms, against the backdrop of a gradual 'drying out' of the environment.

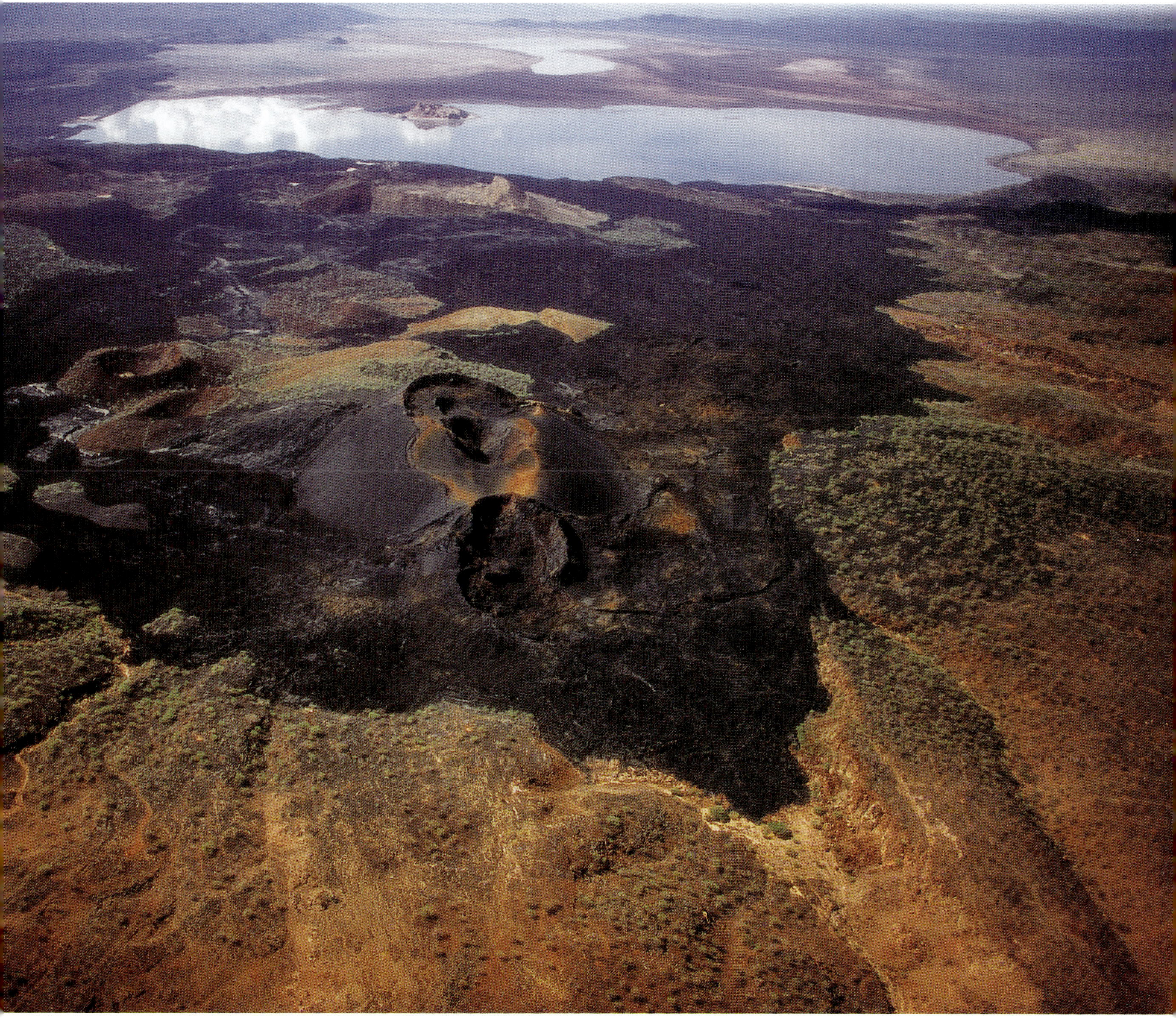

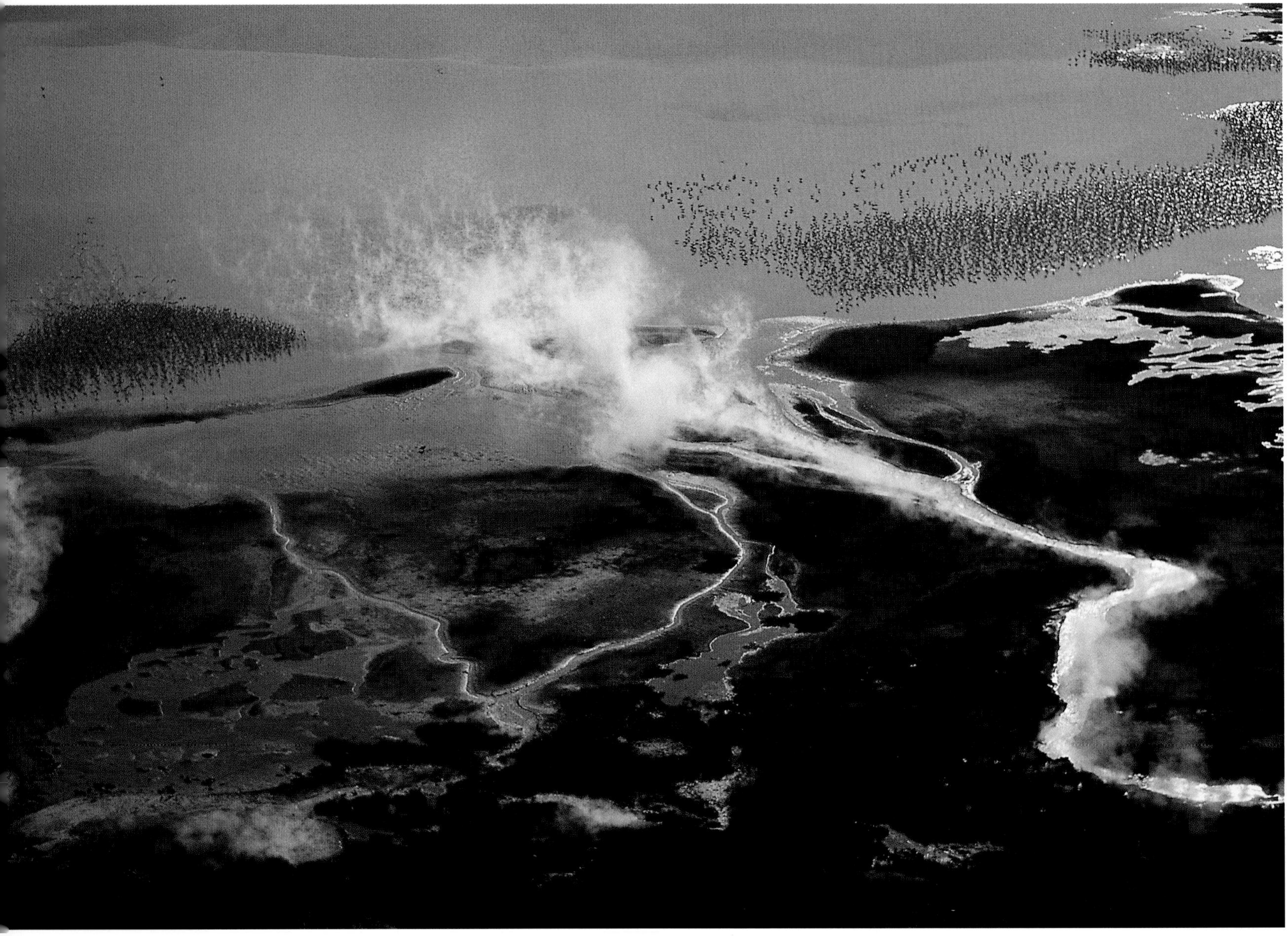

The western shore of Lake Bogoria is studded with geysers, visual testimony to the lively substrate beneath the rift valley floor.

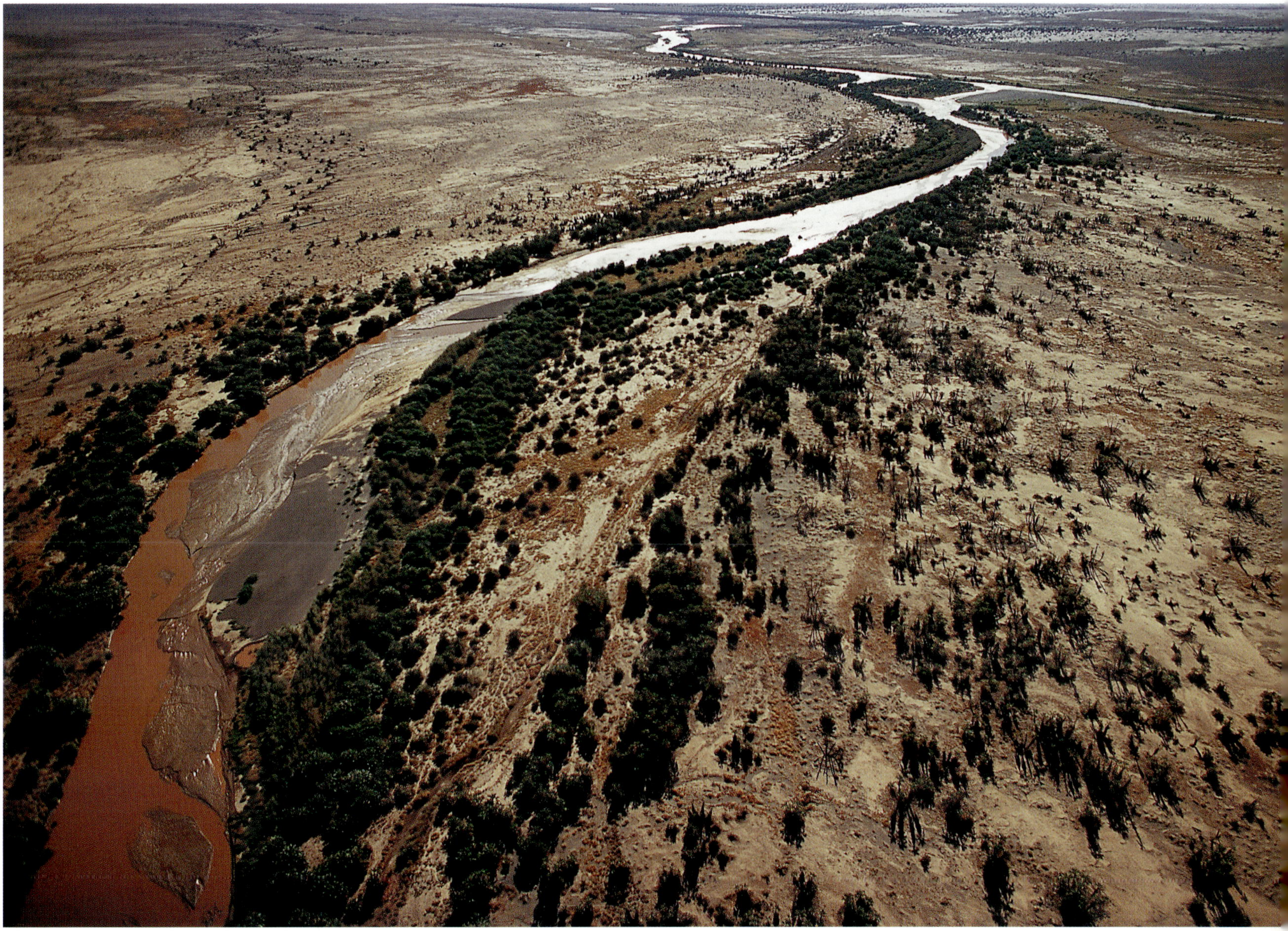

The Suguta River, dry most of the year, snakes along the rift valley floor towards Lake Logipi, south of Lake Turkana.

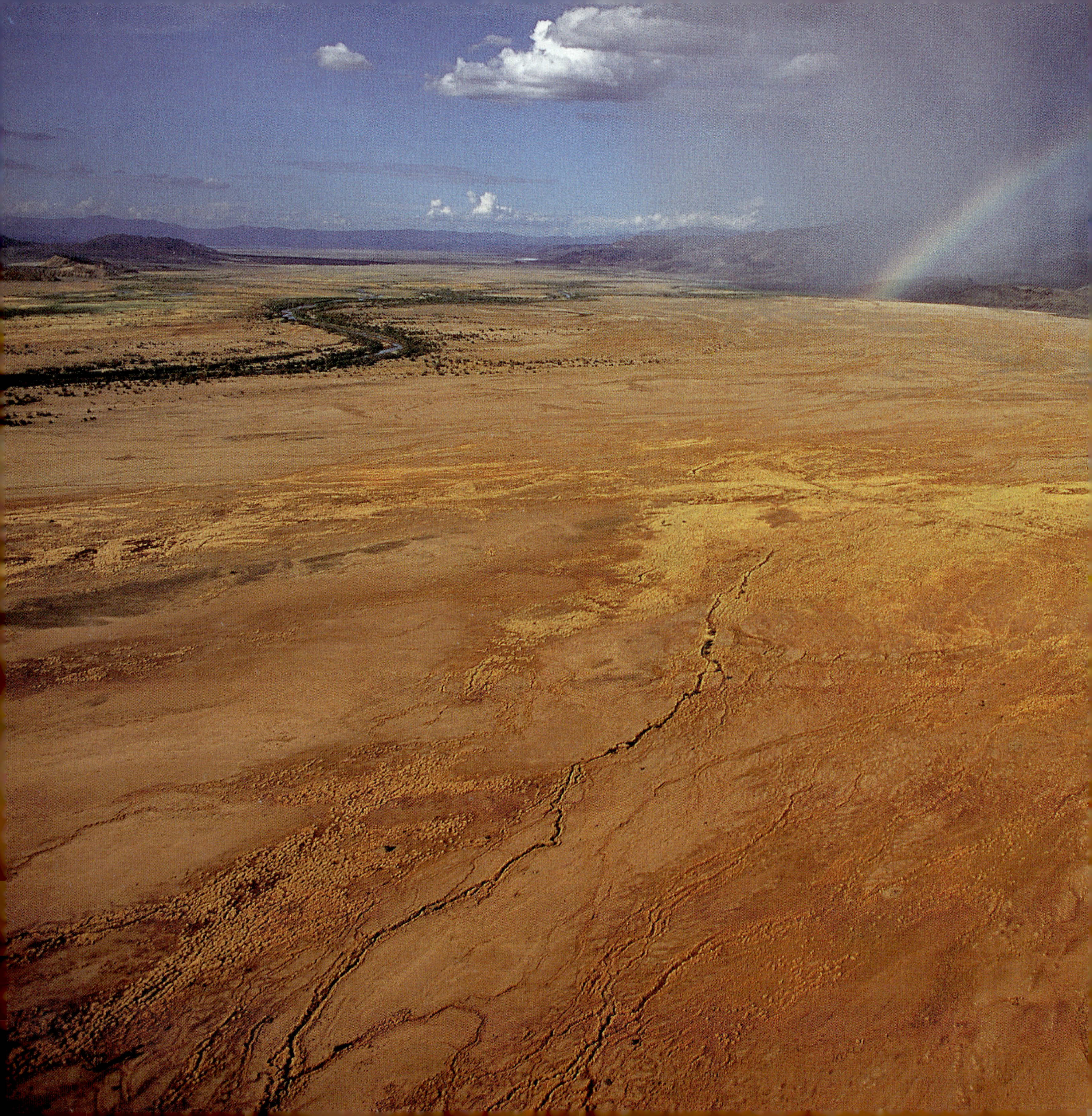

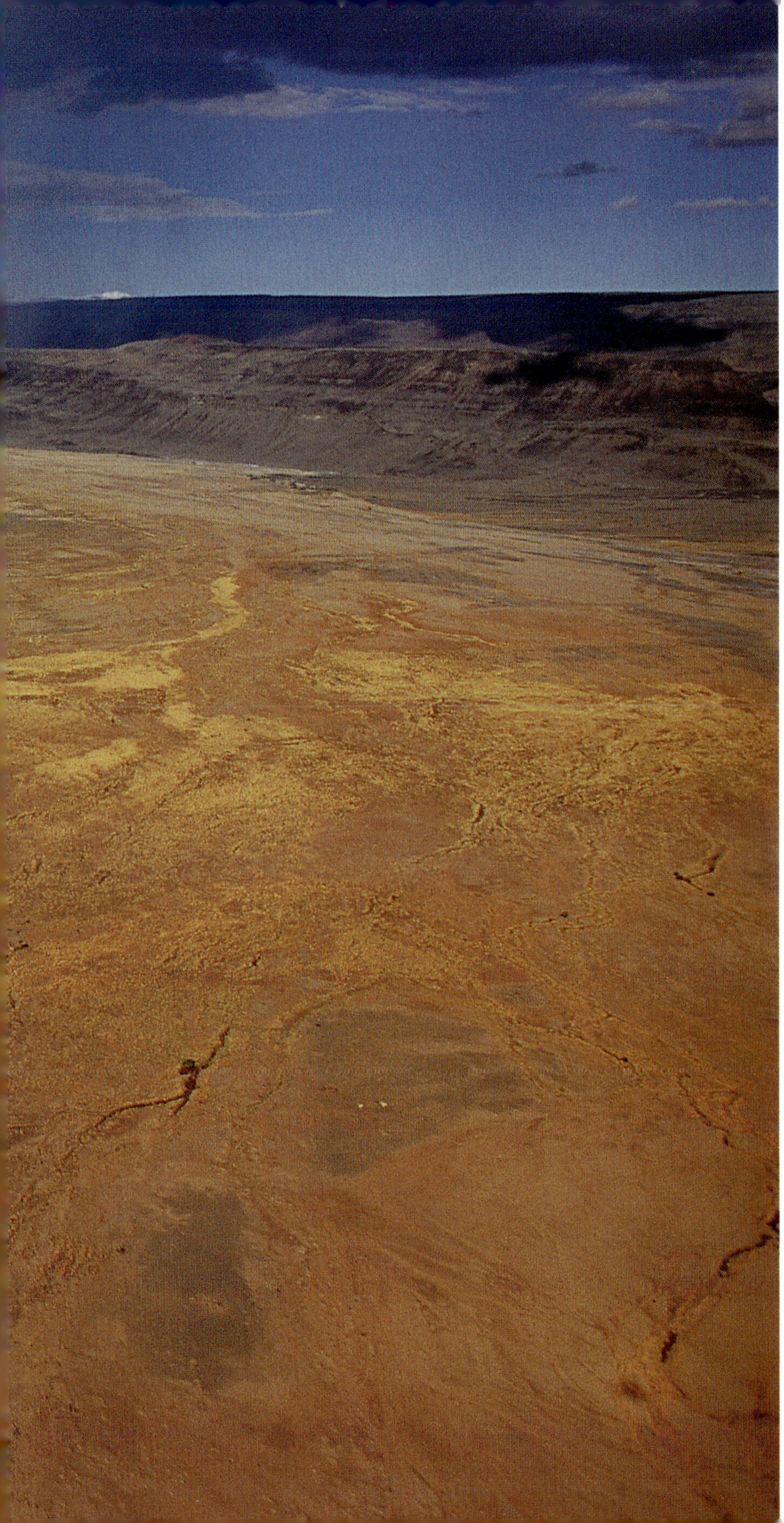

Hominids

In the rift valley landscapes of Kenya and Tanzania, some of the early apes became adapted to living by the shores of lakes, alongside the new animals of the plains. Then, about four million years ago, some of these apes, the hominids, evolved to walk upright, and they also they began to develop bigger brains.

In evolutionary terms, this was a period of rapid and dynamic change – not just for the prehumans, but for a host of other animals as well. Two million years ago, with the emergence at the same time of several protohuman species, the hominids were at their most diverse. Other mammal groups, too, were diversifying, or radiating, into new and different forms. While today there are just three types of pig in East Africa – the warthog, the bushpig and the giant forest hog – there is evidence that at that time there may have been as many as 20 species. The modern lion and spotted hyaena are today the only African carnivores capable of hunting large prey whereas, two million years ago, there may have been 10 species of large carnivore, some much larger than lions. The same pattern is reflected in a host of other animals – elephant, rhinoceros and antelope, as well as primates such as baboons. Each of these groups reached maximum diversity at around the same time.

The hominid brain continued to increase in size and complexity, bringing with it the ability to adapt to different habitats. Eventually, one species evolved a brain so large that any further increase would have required a protective skull too large to emerge through the birth canal of its mother. That species became the ultimate explorer, adapting to live almost anywhere on Earth.

The oven-hot wastes of the Suguta Valley occasionally experience a thunderstorm. The valley was once a southern extension of Lake Turkana, but it is now separated from the lake by a massive volcanic ridge, 914 metres (3 000 feet) high, known as 'The Barrier'.

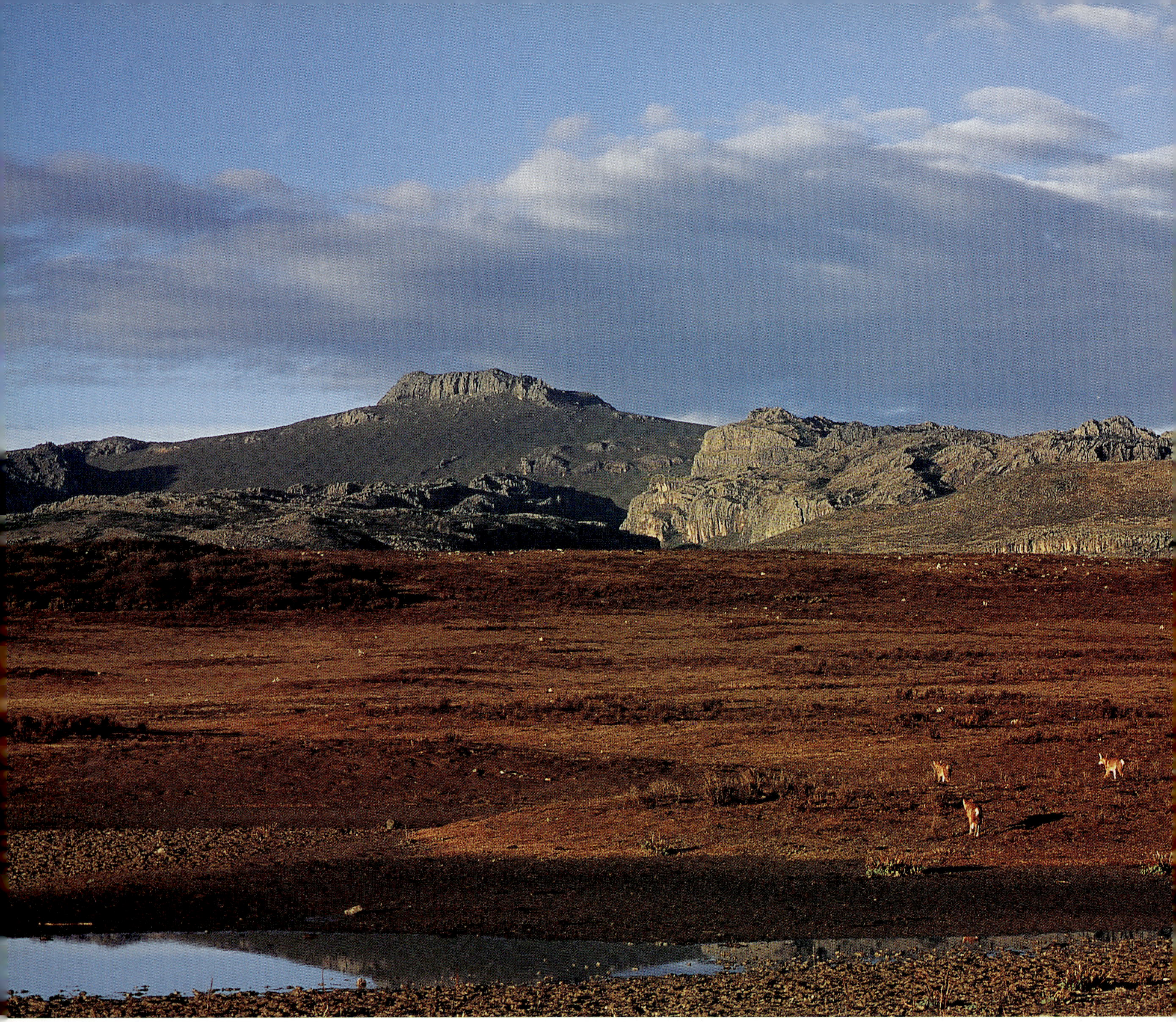

Adult Ethiopian wolves, Canis simensis,
begin their early morning territorial patrol.

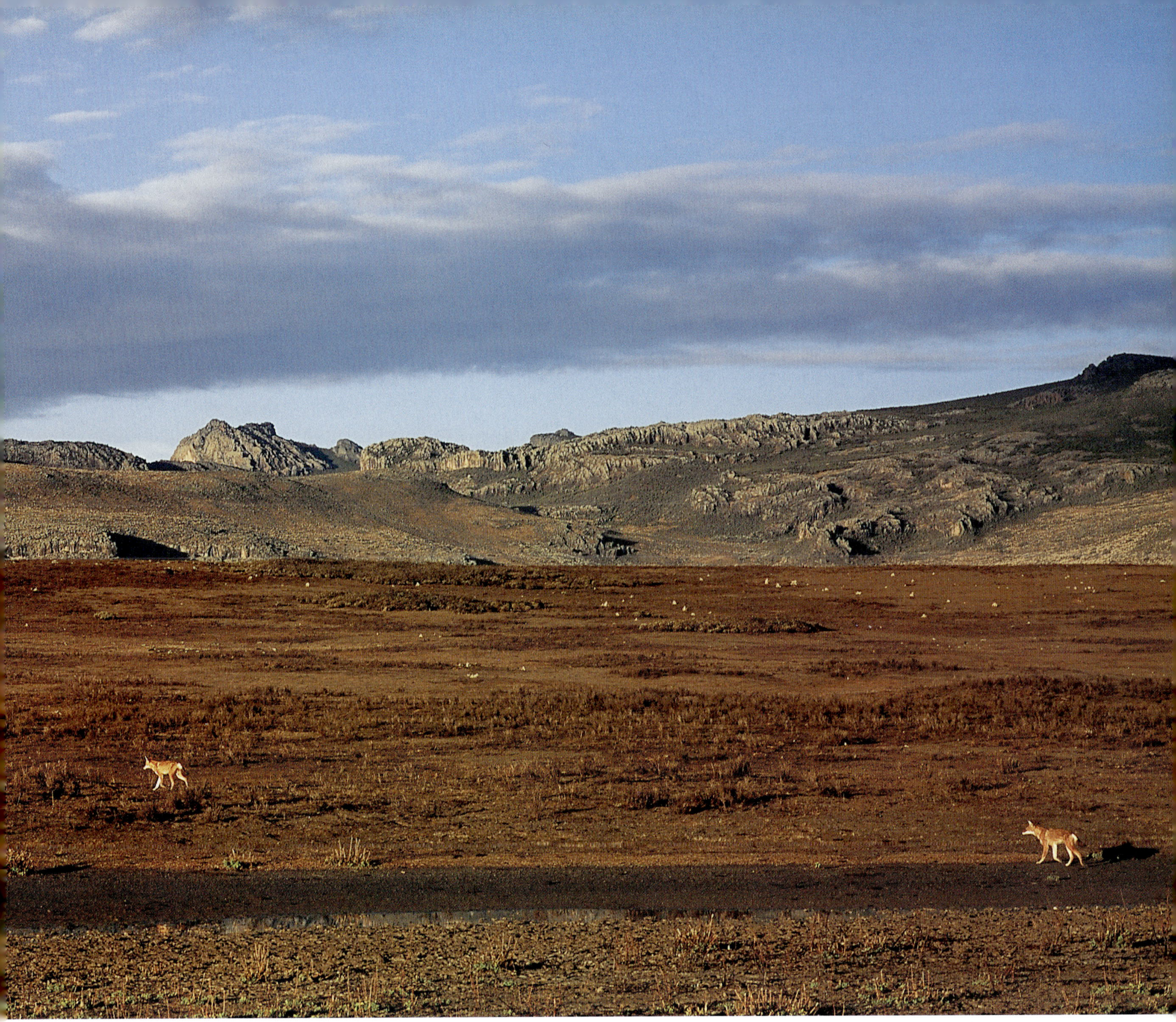

1 Wolves of the highlands

Opposite: The terrain over much of the high Saneti Plateau in Bale Mountains National Park is windswept and austere.
Below: Raptors, like this immature Lanner Falcon, are drawn to the Web Valley by its huge biomass of rodents.
Bottom: A babysitting wolf scans the montane landscape.

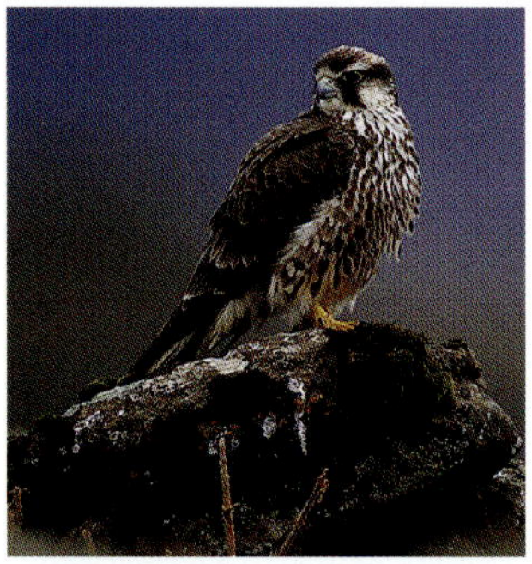

The landscape

Ethiopia's is a remarkable portion of the rift valley, encompassing one half of Africa's land above 2 000 metres (6 560 feet) and four fifths of its land above 3 000 metres (9 840 feet). Many millions of years ago, the Ethiopian Highlands were as flat as the interior of Australia. The violent geological events that formed the rift valley forced up a huge dome, 1 000 kilometres (620 miles) in diameter, and the Ethiopian Highlands – also known as the 'Roof of Africa' – came into being. The highlands are extensive in area and varied in climate; it is not surprising that they support a great many endemic animals – 24 bird species and 23 mammal species.

The most intriguing animals of the highlands are found in Bale Mountains National Park, 400 kilometres (248 miles) southeast of Addis Ababa. The centrepiece of the park is the Saneti Plateau, the world's most extensive tract of Afro-Alpine moorland – a pastel montage of grey and white helichrysum plants, clumped forests of shoulder-high heather, giant lobelias, small, reflective lakes, and lichen-encrusted rocky outcrops.

Web Valley, another spectacular habitat in the Bale Mountains, is a mountain meadow whose herbs and grasses are kept cropped short by the large herds of domestic animals of the Oromo pastoral people. The ground is spongy underfoot, punctured with holes created by the area's extraordinarily dense concentration of rodents.

The bird life in the Bale Mountains is complex: migrant raptors co-exist with local raptors; other migratory species and breeding birds are found alongside endemics. Mountain mammals include the endemic mountain nyala and eight species of endemic rodents, among them the giant mole rat. At 2 500 kilograms (5 525 pounds) per square kilometre, the biomass of rodents in some areas of Web Valley matches that of the large ungulates of the Serengeti-Mara ecosystem. But the most remarkable mammal is the one that preys on the rodents – the Ethiopian wolf.

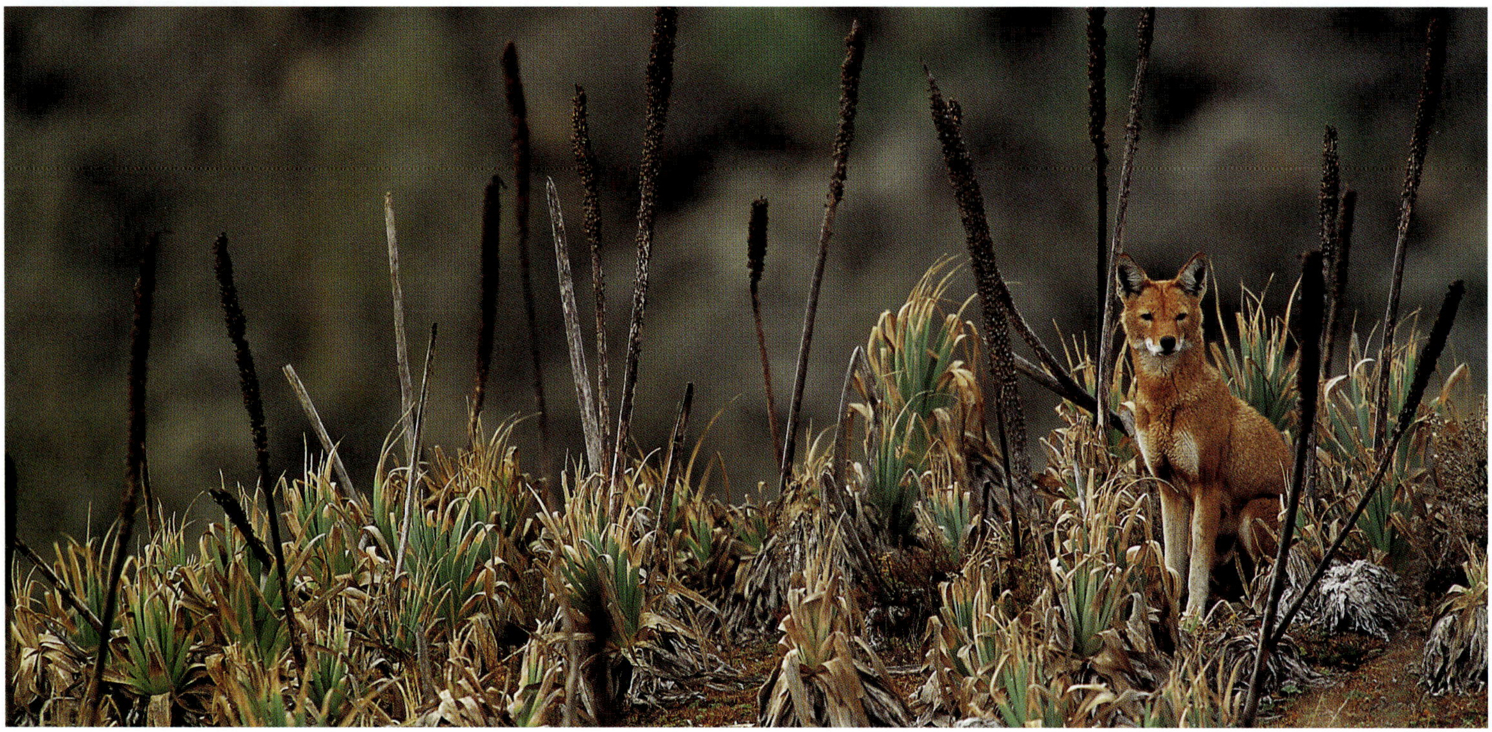

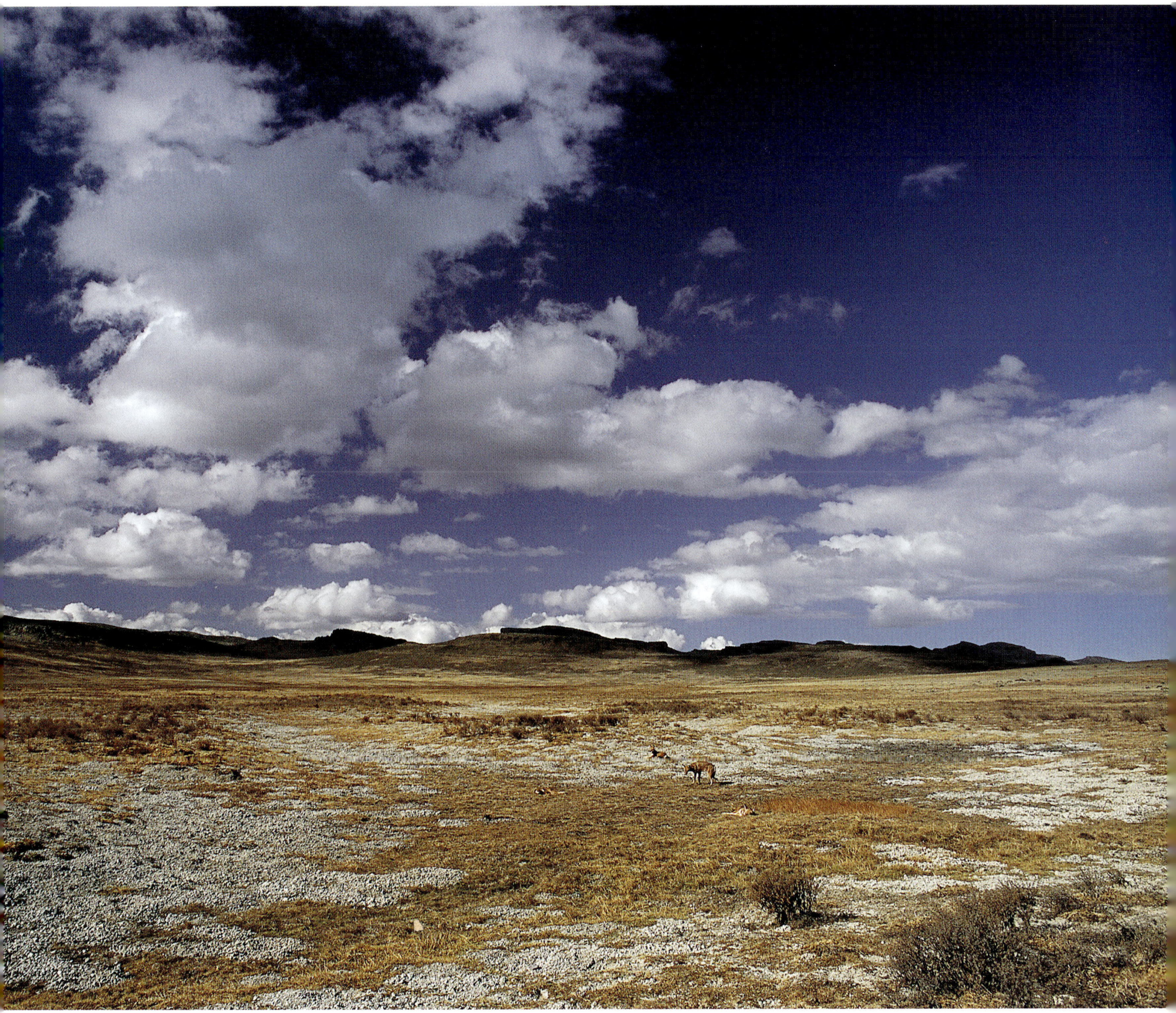

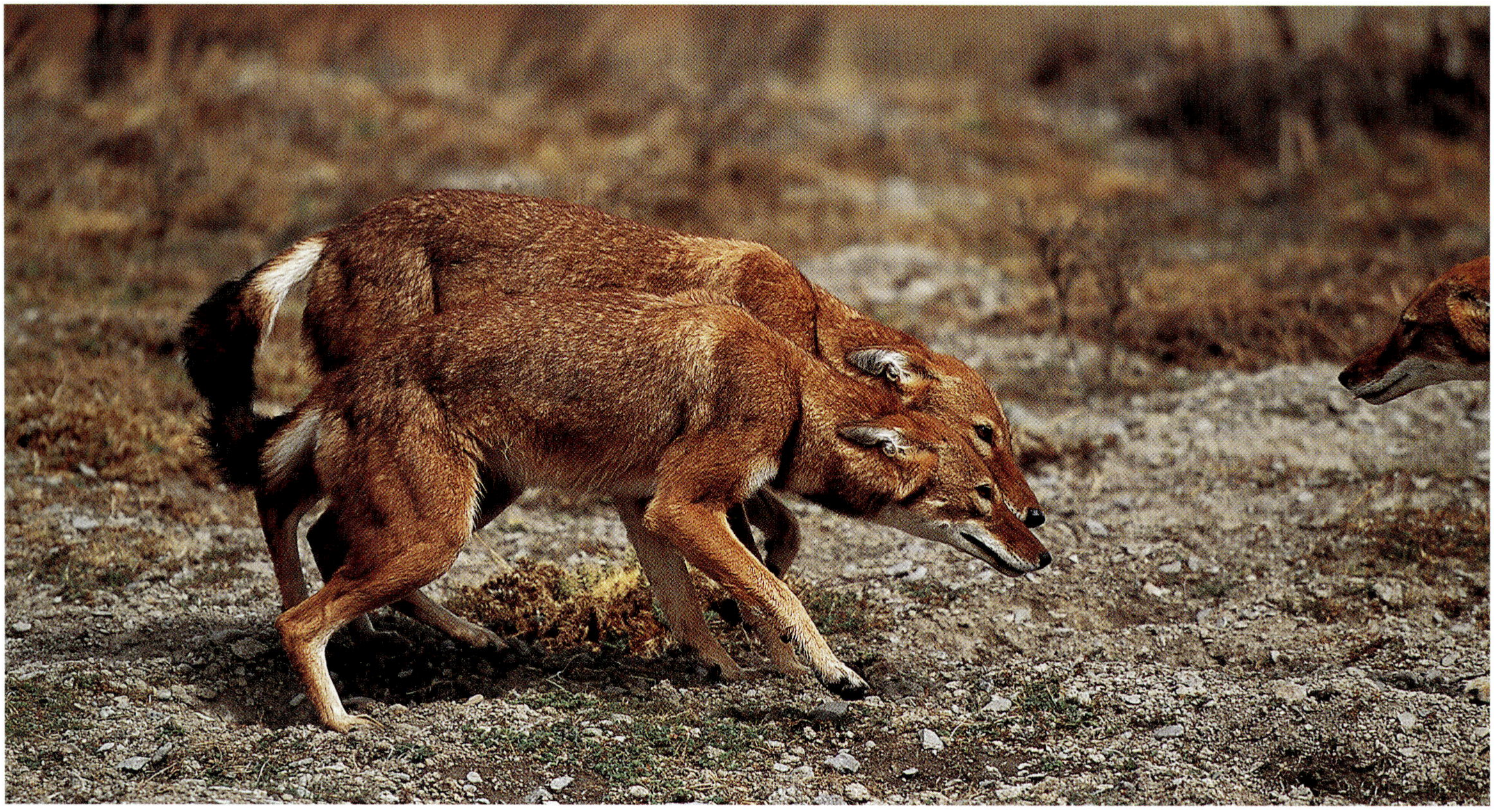

Above: *Ears laid back, two crouching wolves signal submission. A strict hierarchy rules in the wolf world; subordinates regularly signal submission when approaching a higher-ranking wolf.*
Opposite: *An adult and pup keenly watch the den entrance for the remaining pups to emerge.*

The Ethiopian wolf

The German naturalist, Eduard Rüppell, first described the Ethiopian wolf in 1835. Then, around the turn of the twentieth century, a German ornithologist, Carlo von Erlanger, visited the region, followed in the late 1950s by the Finnish geographer, Helmer Smels. These early naturalists mistook the wolves for jackals or foxes. The animals had been observed to forage alone and to bark like jackals, while their orange-red coats were reminiscent of those of foxes – but their genetic profiles subsequently revealed their closest relatives to be the grey wolves of Eurasia.

Wolf origins

The wolves are, in fact, a leftover ice-age species in a chilly remnant of Africa's ice-age world. Some 100 000 years ago, during a global ice age, the Ethiopian Highlands were connected to the more temperate biomes to the north. Glaciers covered the peaks and plateaus of the Ethiopian Highlands, and broad vegetation belts persisted at between 1 000 and 1 500 metres (3 280 and 4 920 feet) lower than their present altitude. A small number of grey wolf-like ancestors ventured into this suitably wintry land from Eurasia. They would have been unable to penetrate further south into Africa as the habitat was unsuitable and they would certainly have encountered fierce competition from wild dogs, hyaenas and lions. When the climate warmed, the broad belt of sub-Alpine vegetation on which the isolated wolves depended contracted to higher altitudes, and the wolves also moved up. Marooned in the process on the 'Roof of Africa', they evolved into the Ethiopian wolves of today.

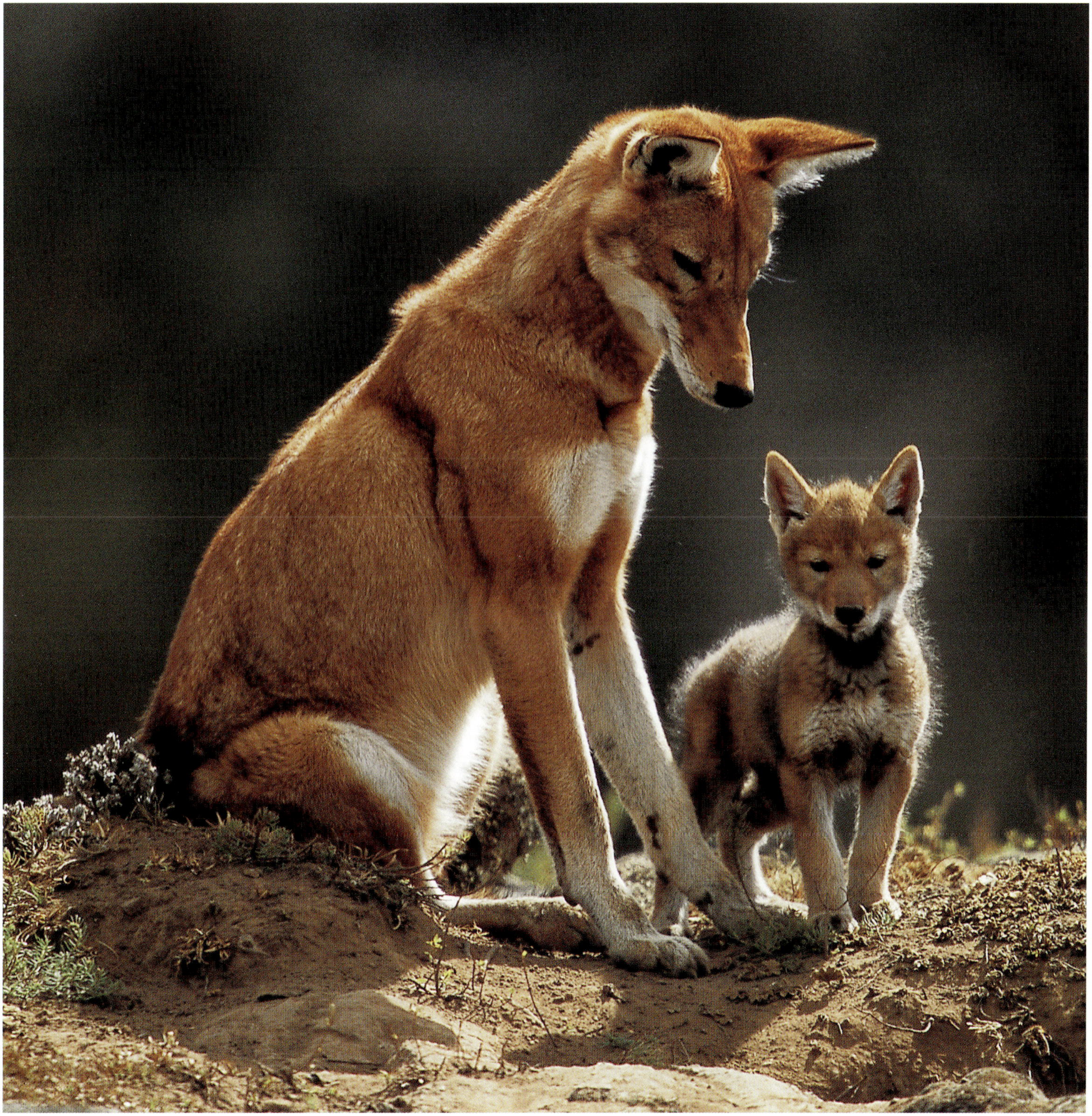

Wolf behaviour

Ethiopian wolves are social animals, living in cohesive units of up to 13 adults and yearlings. Each pack defends a territory, the size of which depends on the density of available prey. Adult males stay in their natal range, whereas two thirds of the females disperse to seek breeding opportunities elsewhere, sustaining higher mortality rates in the process. A wolf pack is held together by a dominance hierarchy. The relative positions in the hierarchy are continually reinforced through body language and mock sparring. This strategy avoids the dangers of real fights and quarrels.

At the head of the pack is an alpha breeding pair, and within the pack, normally only the alpha female gives birth. On the rare occasion when a subordinate also gives birth, the alpha female kills the subordinate's pups. When she is on the pack's territory, only the alpha male mates with her. When foraging alone in the mating season, however, she may sneak into neighbouring territories and mate with any male who comes along.

The pack uses a den only during the breeding season, at which time it is a focal point; pack members come and go during the daytime, but sleep close to the den at night. Up to six pups are born, and during the first few weeks the mother changes the den two or three times. When the pups have been weaned, they are watched over by a 'babysitter' while the parents and the rest of the pack are on territorial patrol. Every pack member helps in raising the pups by protecting them and feeding them with regurgitated food. This behaviour is puzzling, since the carers do not necessarily share genes with the pups they are helping to raise.

Returning from an early morning territorial patrol, a mother is mobbed by hungry pups, eager to suckle.

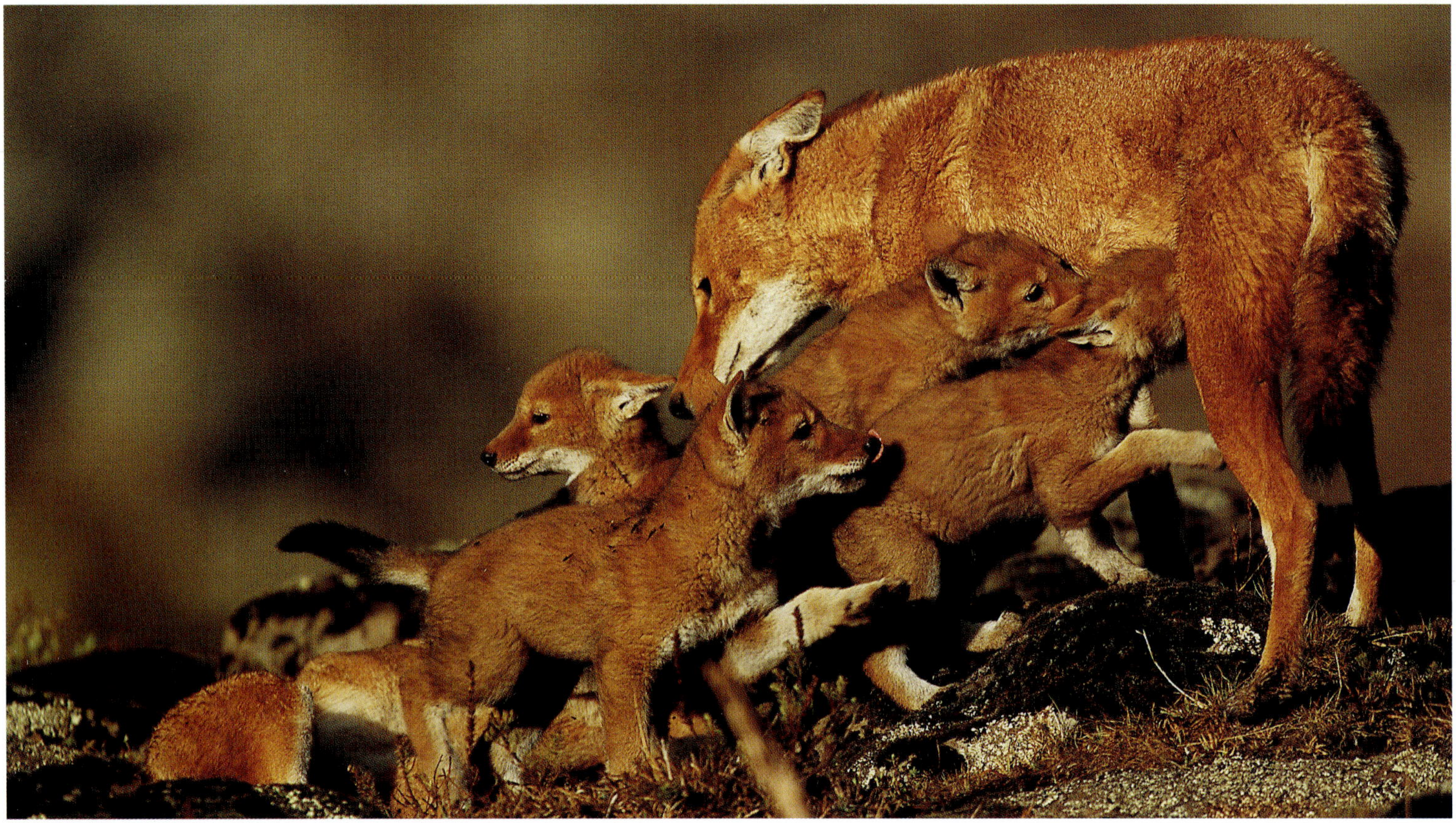

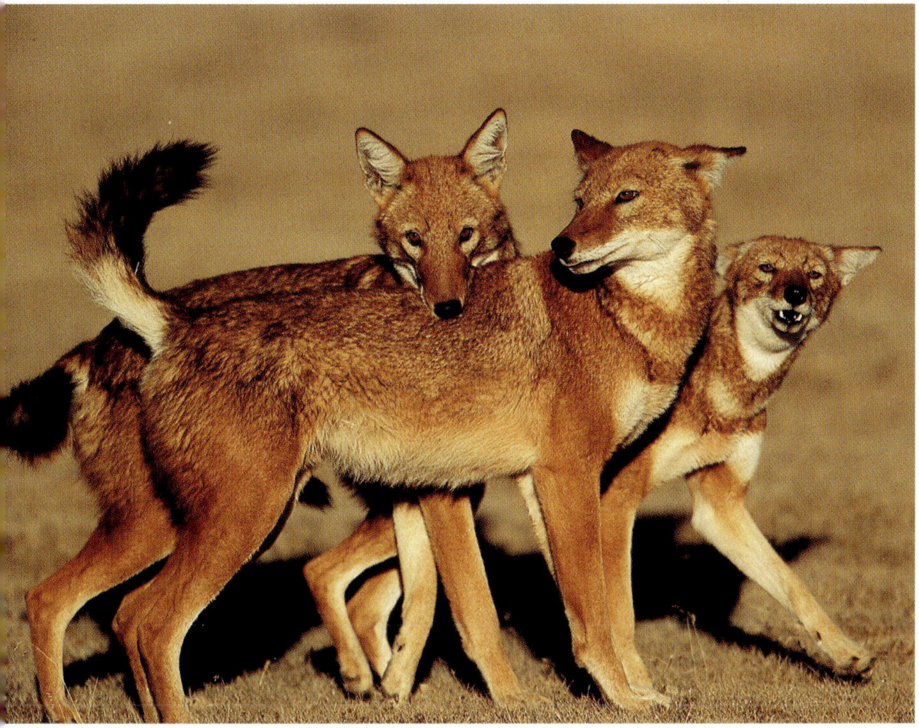

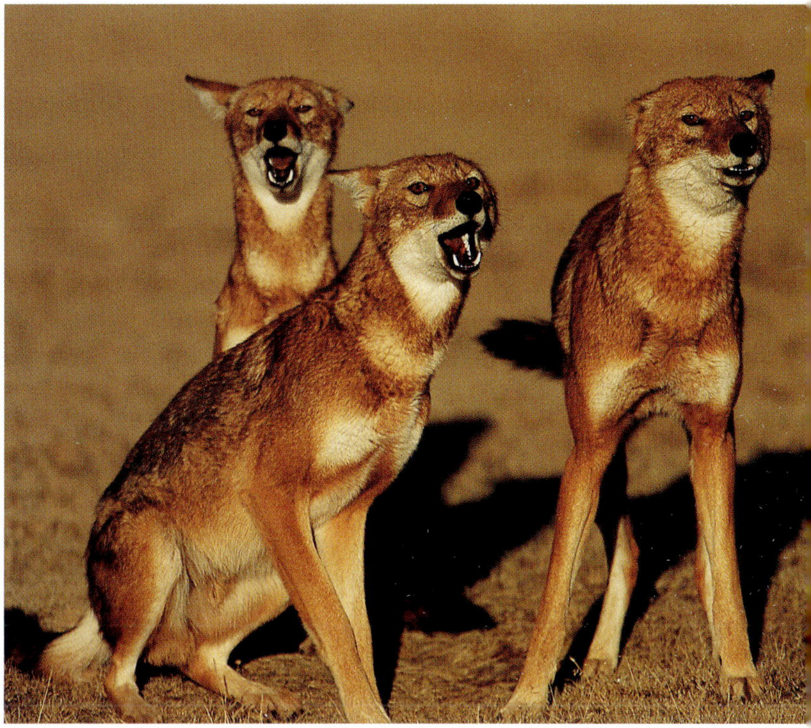

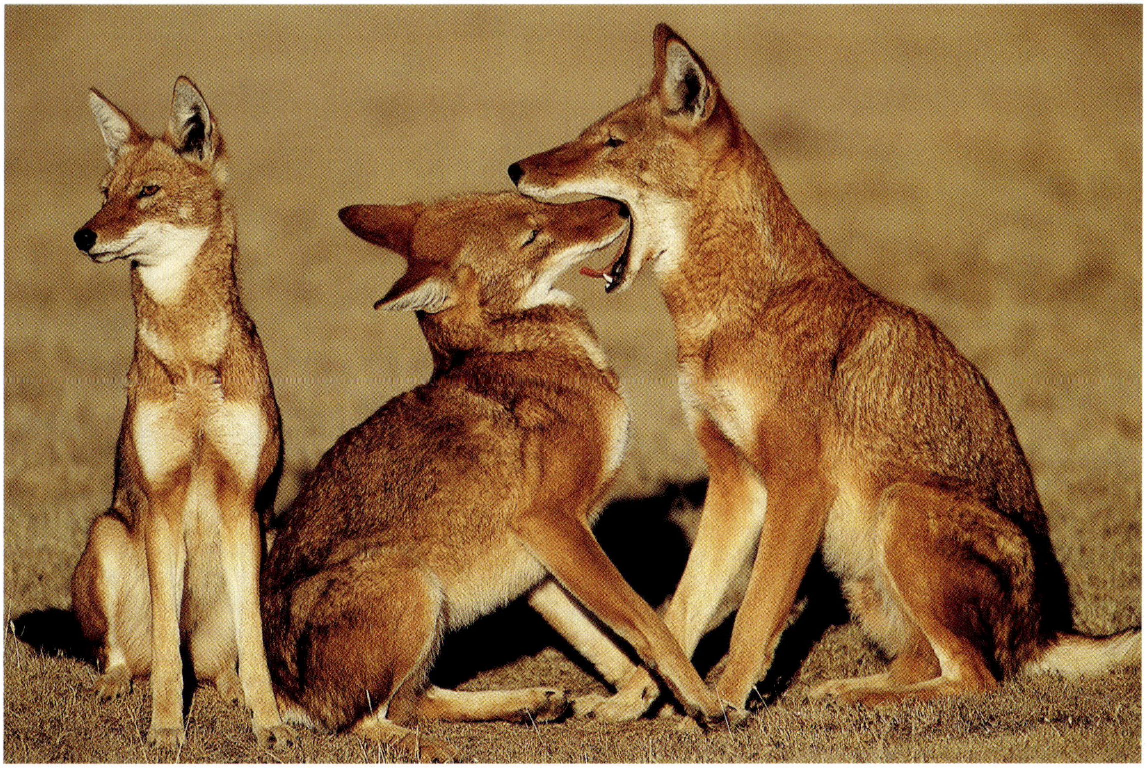

Above left: Sniffing, snapping, yipping, whimpering – wolves in a pack greet one another in a morning ritual that strengthens the solidarity of the pack while reinforcing its pecking order.

Above: Vocal exchanges between groups of wolves enable the pack as a whole to remain in contact.

Left: Greeting after a brief separation, two wolves play: one wolf puts its muzzle inside the mouth of another. Jackals, foxes and hyaenas occasionally play-bond in this way. This behaviour is likely to be a signal of trust.

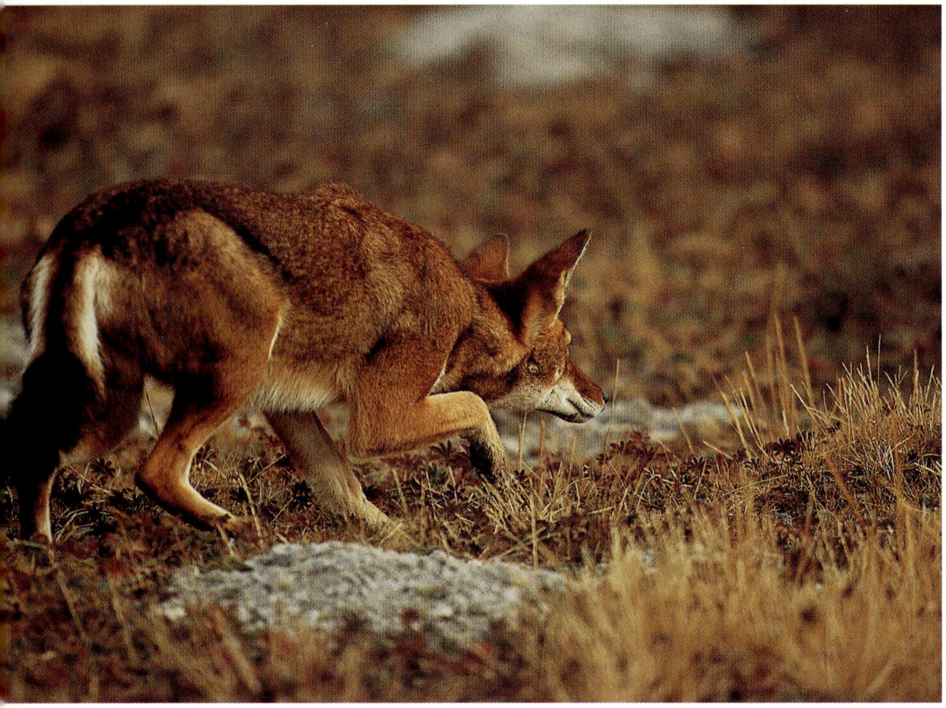

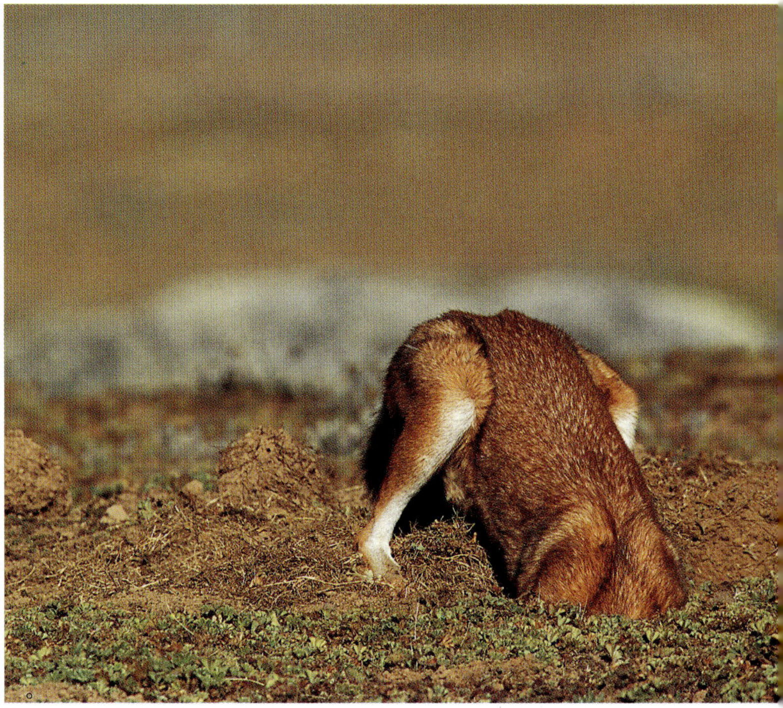

Above: *Ethiopian wolves regularly stalk to get close to targeted prey without detection. They may stay motionless for a few minutes before pouncing on the prey animal.*
Above right: *Having seen a rat disappear down a hole, this wolf made determined efforts to prise it out, this time to no avail, although the technique often does result in a catch.*

Well-fed pups are extremely playful; in a recurring dominance game, one pup tries to assert itself by mounting the other. In this way, hierarchy among pups is established at an early age. The pups are also highly curious, with an uncontrollable urge to explore. Pack members, particularly subordinates, spend a great deal of time playing with the pups.

The high plateaus of the Ethiopian Highlands can be bitingly cold: in the dry season the temperature can drop to -15 °C at night; yet the wolves do not make burrows, but rather sleep out in the freezing open air. When the sun streams across the plateau, they wake up, one by one, and go through a boisterous greeting ceremony. Then the pack sets off at a rapid trot to patrol its territory. The wolves move around the area, investigating all the smells they encounter, and liberally depositing their scents at strategic spots on the territory boundary. When wolves from two different packs meet at the common border, tension fills the air. After vigorous vocalising and posturing, the pack that is outnumbered retreats into its territory; with the dispute settled, both packs continue their separate ways.

Once the patrol is over, the pack splits up. Each wolf forages alone, feeding primarily on rodents, particularly rats. With its sensitive ears, slender muzzle and sharp, outward-pointing front teeth, it is a specialist rodent catcher. An excellent hunter, it catches rodents in a number of ways. The most common technique is stalking: the wolf, with consummate patience, takes small steps, sometimes in a crouched position, freezing, stalking and freezing again, for up to 10 minutes at a time if necessary, until eventually it pounces on the unsuspecting prey. Sometimes, it will dig away half a metre of soil to get at a nest of young rodents; occasionally, it pads the ground to flush out a rodent, leaping on its prey as the rodent emerges from its burrow. Ethiopian wolves will occasionally revert to ancestral type and group together to hunt large prey – a young mountain nyala or a reedbuck. These attempts usually fail, but even their close relatives, the more typically wolf-like Eurasian grey wolves, are successful only once in every 10 attempts.

Sociality

During the first major study of the species in the Bale Mountains, the strong social nature of the Ethiopian wolf puzzled researchers. Since the animals normally hunted alone, sociality did not appear to make sense. Why would these wolves live in packs if they were solitary hunters and did not share their meals? What advantage might a wolf derive from helping to raise the offspring of another, unrelated wolf? An answer emerged only after the researchers had counted the rodents available to each pack in its territory: it then became apparent that communal living provides the wolves with a means of securing territories that are large enough for their populations to survive. Together, a pack can defend an area with sufficient rodents to feed each individual and the communal pups.

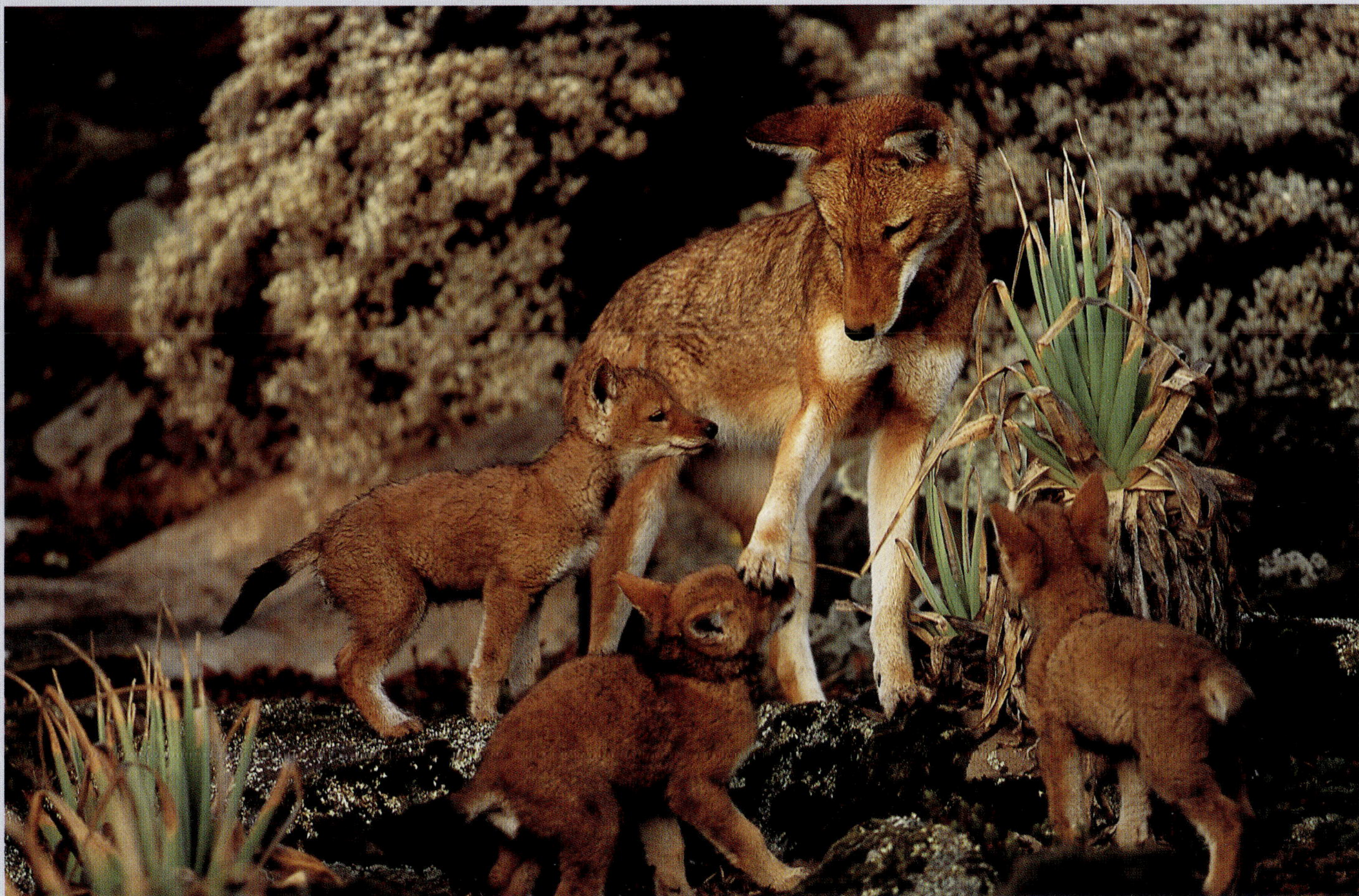

The Ethiopian wolf is an extreme example of an animal in isolation – the very factor that has enabled it to evolve adaptations to its unique environment. As a consequence, although indubitably a wolf, it exhibits some behaviour, especially its feeding strategy, that sets it apart from its Eurasian cousins. The only wolf in Africa, it has the dubious distinction of being the most rare canid species in the world.

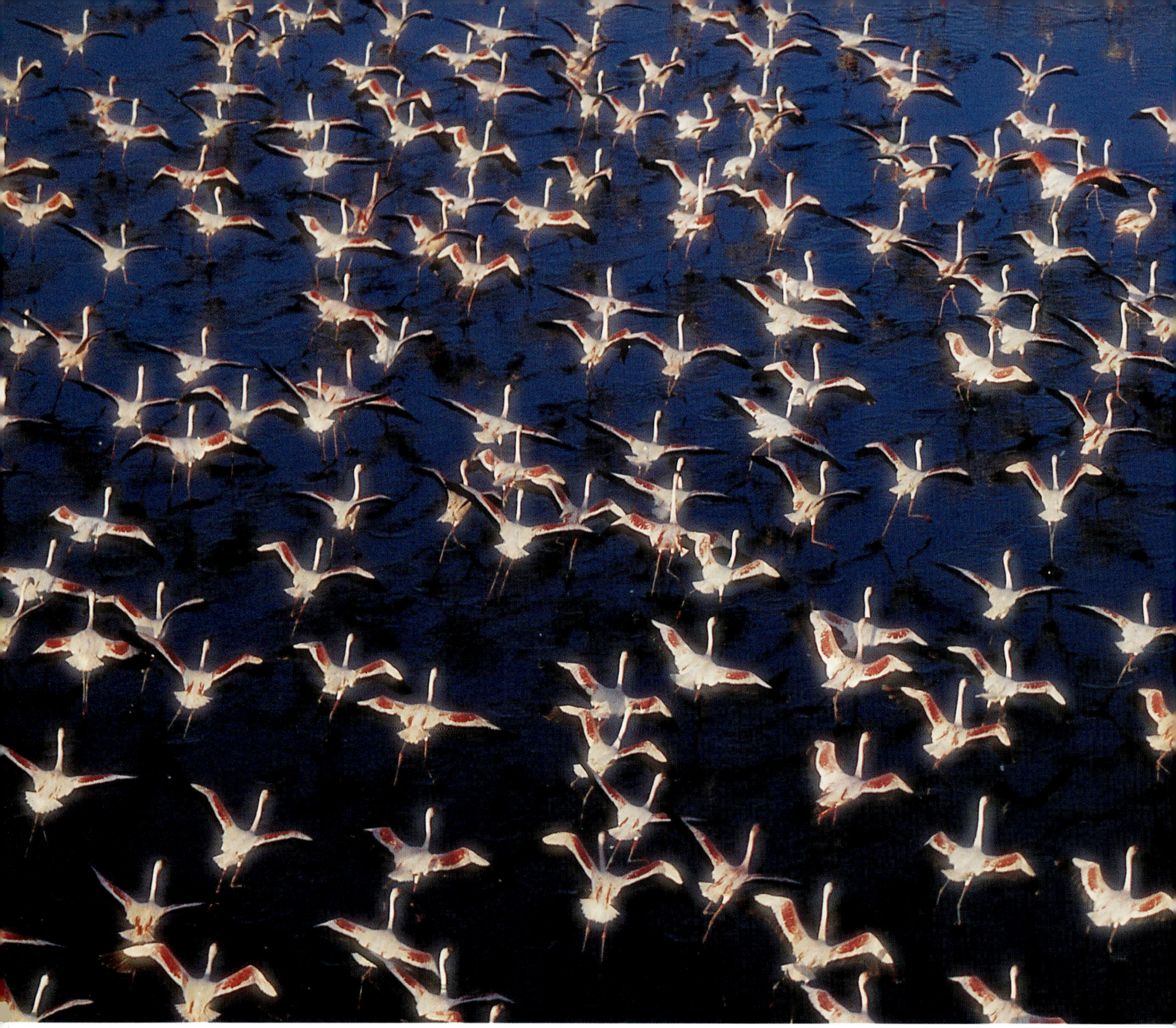

A great cloud of Lesser Flamingos, Phoenicopterus minor, *takes to the air, wings flapping. Once airborne, they will stretch out their necks to counterbalance their long legs.*

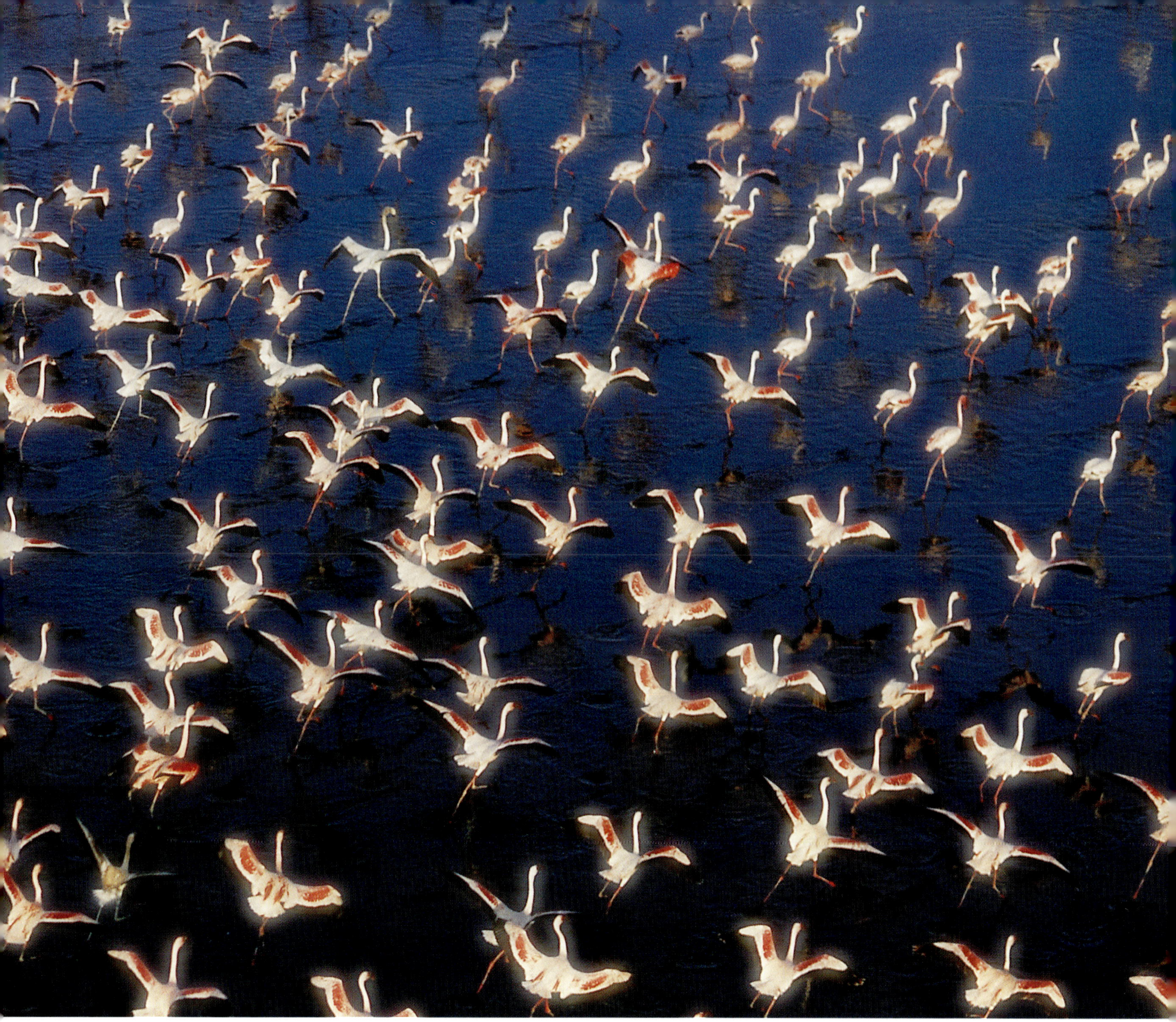

2 Flamingos of the soda lakes

The lake environment

Twenty-five million years ago, there were no lakes in Africa; but then violent geological forces reshaped the continent to form the Great Rift Valley, creating many places where water could settle. The life drawn to the shores of such lakes was preserved as fossils by the favourable rift geology; volcanic deposits, built up in and around the lakes, preserved the remains of animals that perished at their edges, to be discovered, several million years later, by palaeontologists when subsequent faulting and erosion laid them bare.

Soda lakes

Not all lakes, however, attracted such diverse life; during the formation of a special group of lakes, much of the ground in the rift valley was covered in volcanic ash, rich in sodium carbonate (known domestically as 'washing soda'). Pure water, raining or condensing on the volcanic ranges, would flow down the slopes – sometimes underground, surfacing as a spring when it reached the junction of two slopes – and gather volume to become a river. While percolating through the soil and flowing over rocks, the water absorbed and transported volcanic chemicals and mineral salts into the lakes. In the closed, unconnected lake basins without water outlets, the dissolved salts accumulated and were concentrated through evaporation in the lakes, leaving behind a sodium compound also found in sodium carbonate, sodium bicarbonate (baking soda) and sodium hydroxide (caustic soda), imparting a bitter, soapy taste to the water.

Such bodies of water are known as soda lakes. Concentrations vary from lake to lake, but water from a highly concentrated lake is corrosive to the tongue. The concentration can become so high as to leave the lower limbs of a mammal standing in the lake badly scalded or seared. These lakes are inhospitable to most life forms – yet in the rift valley, those most ancient of birds, the flamingos, thrive on them. Clearly they have tapped into a sustaining resource.

Below: *Gawky and vulnerable, Lesser Flamingos run through the shallows and beat their wings for take-off.*
Opposite: *An aerial view of a shore at Lake Natron, where flamingos breed. Sometimes the lake looks like the surface of another planet, an acrid chemical mix where it might seem surprising to find life.*

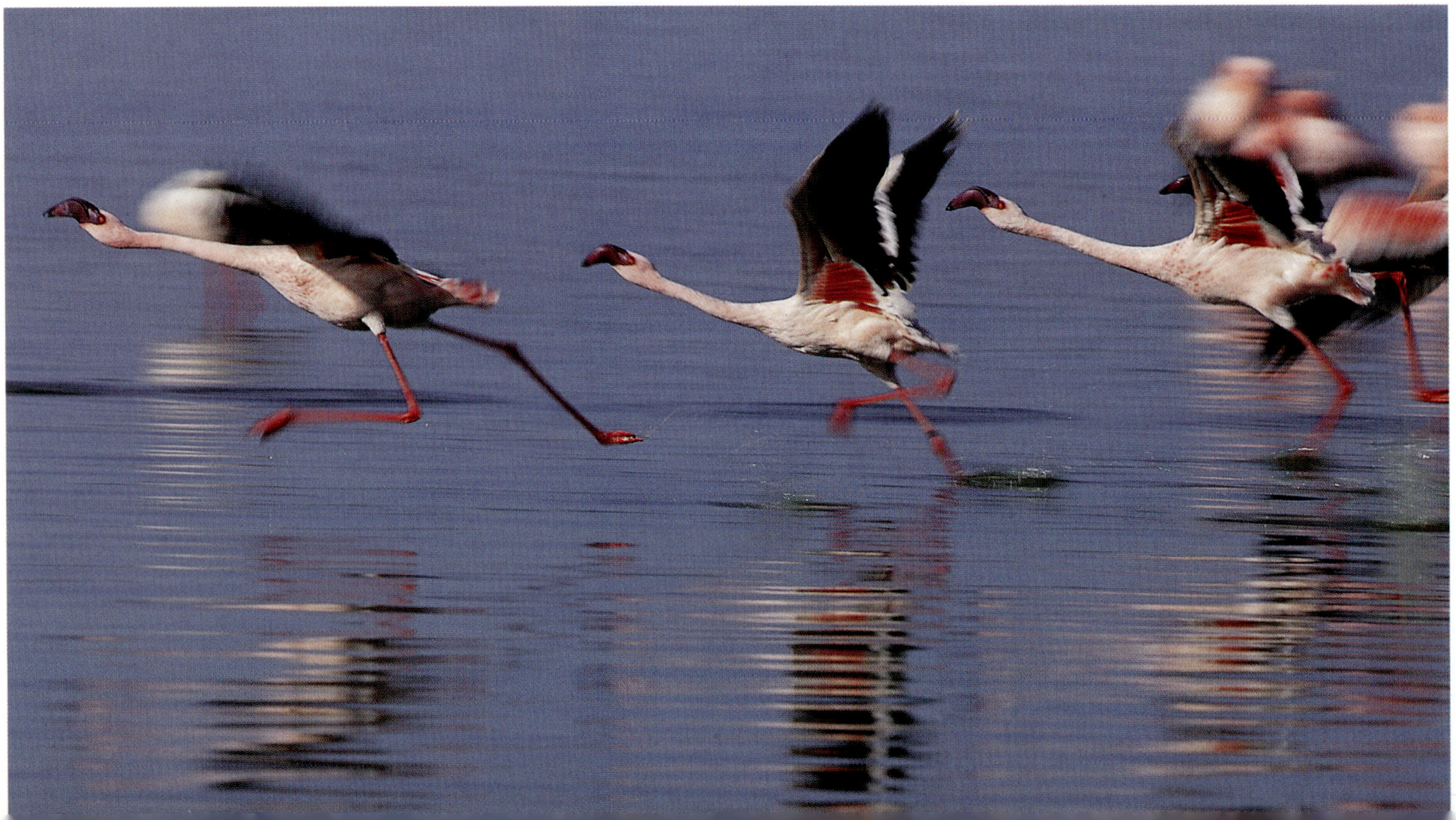

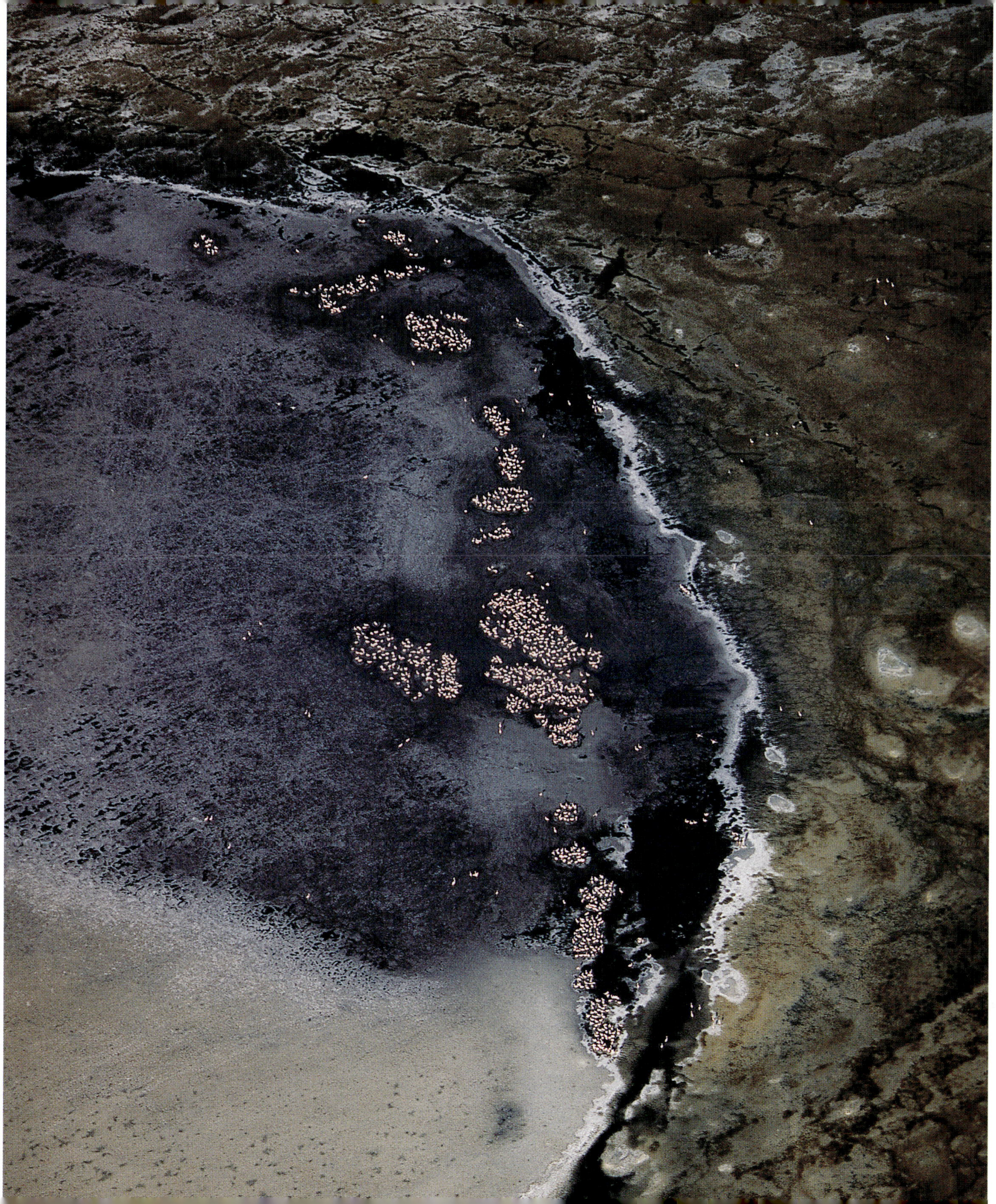

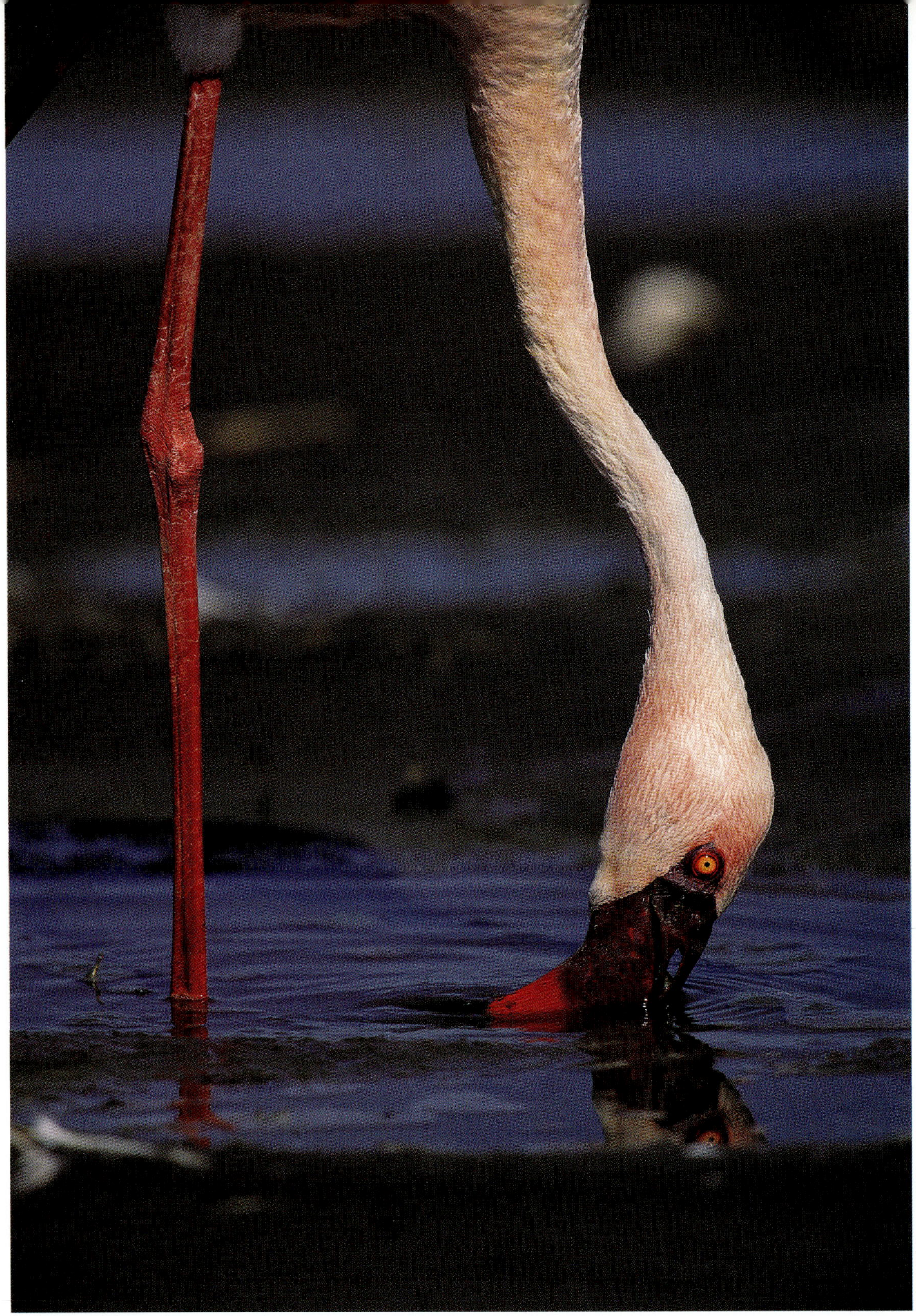

With its head in the characteristic upside-down position, a Lesser Flamingo sucks and sifts algae from the calm soda shallows. Its bill is heavily arched to maximise the quantity of food it can 'hoover' up, using the hair-like lamellae along and inside the mandibles to trap great quantities of microscopic plant food.

Although the lakes of the rift valley, with their seething chemicals, are alien to most familiar life forms, a few plant species have adapted to cope with the challenging conditions. Of these, the most abundant are the many species of microscopic single-celled algae, visible only when magnified a thousand times. Some live as single cells, while others clump together. All algae produce their own food through photosynthesis, as they lie suspended in the water practising their ancient lifestyle.

Algae are prodigious breeders – reproduction may be sexual or asexual. In the thick, slimy green waters of Lake Nakuru, an object becomes invisible just 15 centimetres (6 inches) below the surface, obscured by a mass of blue-green algae, *Spirulina platensis*, that turns the lake into a murky soup. *Spirulina*, in particular, thrives on the combination of the carbonate- and phosphate-rich waters, the high temperatures and the plentiful sunlight characteristic of some rift valley lakes, and provides the key ingredient in the success of the Lesser Flamingos. These birds have evolved to exploit harvests of *Spirulina*; the vast quantities of this aquatic plant have ensured that flamingos are present in huge numbers – flocks on a bountiful lake can number two million, by far the largest gathering of any inland water bird in the world. Flamingos appear to have claimed exclusive rights to these strange lakes with their bumper crops.

Lake Bogoria lies to one side of the magnificent Laikipia Escarpment, and is fed partly by hot springs that bubble from the rocks on its shore.

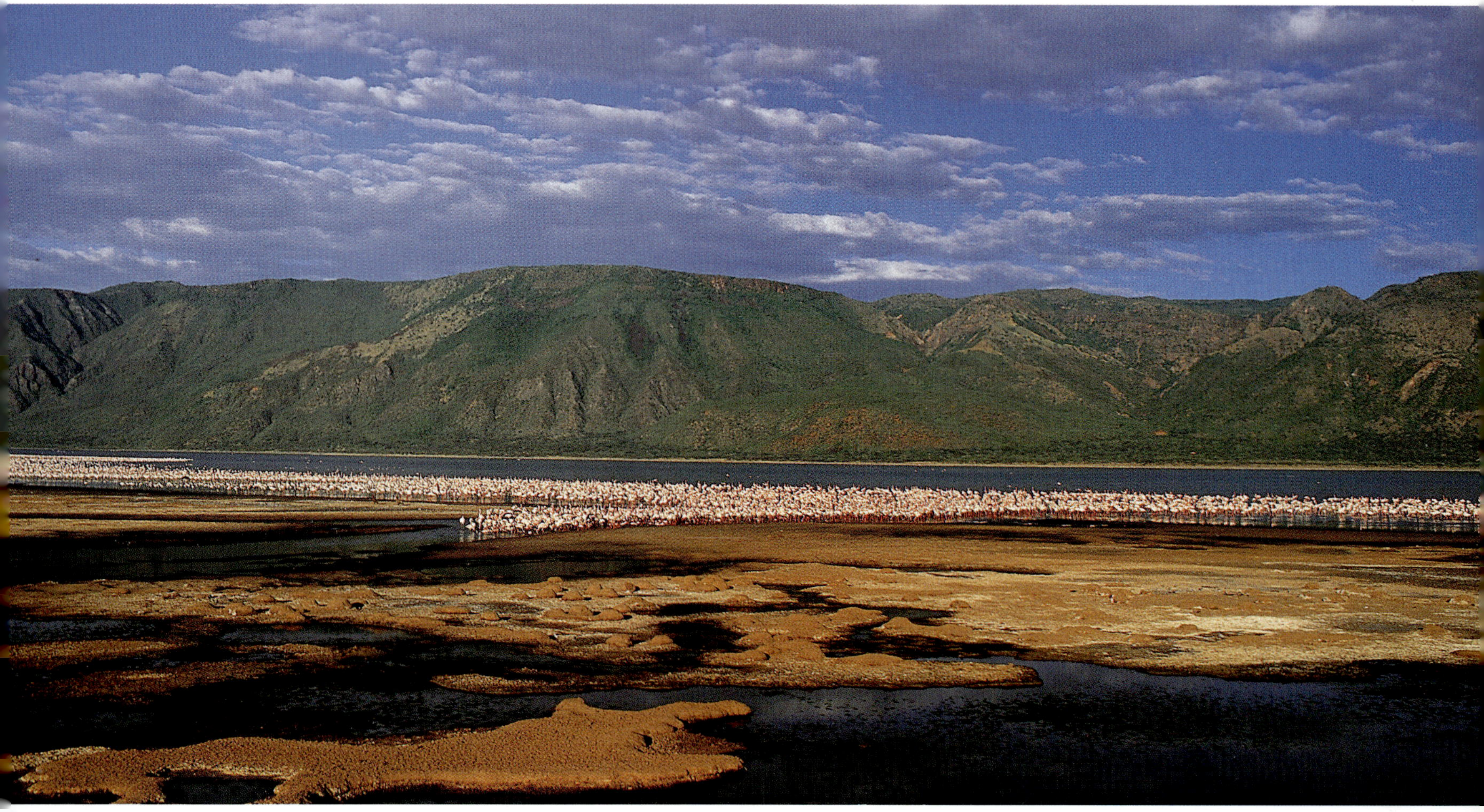

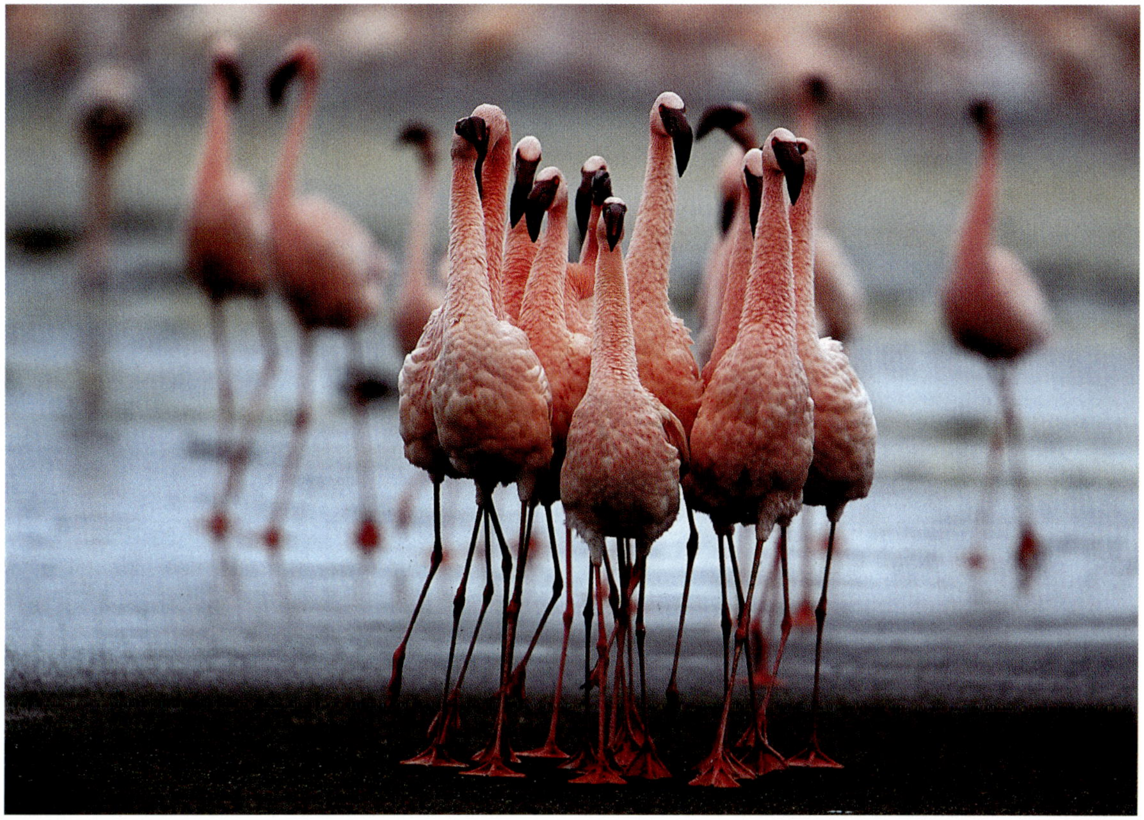

Looking for mates, male flamingos perform a courtship display on the shores of Lake Nakuru.

Flamingo adaptations to lake life

The hardy flamingo, able to withstand the inhospitable soda lake habitat and to exploit a food source resembling 'dissolved grass', is a marvellously adapted specialist. It has long, entirely featherless legs and thighs, with tough outer coverings, which allow it to wade in caustic, corrosive water up to a metre deep (though an accidental cut on the leg from the sharp lava rock could result in a wound that will not heal and might lead to premature death). In shallower water, a flamingo's long neck extends downward to reach the water's surface; its head can reach its webbed feet, which are perfectly adapted for walking on the uneven lakebed. Wedged into this small head is a grotesque bill – surely its most fascinating adaptation. A flamingo may often be seen harvesting algae by 'grazing' in the upper two centimetres (one inch) of water, walking with its head lowered and its bill submerged upside down. The bill is heavily arched so that great quantities of algae can be sucked up over the fine hairs inside its mandibles, which filter out the rich food. The bird copes with intense heat while it feeds, bathes and preens, and can breed even in exposed areas where temperatures exceeding 50 °C are normal.

With these adaptations, the flamingo looks like a monstrously stretched goose, with webbed feet and waterproof feathers. Like other waterfowl, it swims and upends, honks, filter-feeds and flies in flocks, and uses similar drawing-in movements of the head to build a nest; its downy chicks resemble goslings; the genera of lice that live in flamingo feathers are also found in duck feathers. Could it be an eccentrically adapted distant relative of those other waterfowl – the ducks, geese and swans?

Lesser and Greater Flamingos

The Lesser Flamingo, *Phoeniconaias minor*, or 'smaller crimson water nymph', obtains its exquisite colour from the carotenoid pigments in the algae, which are broken down by enzymes in the bird's liver to become feather pigments. The pink is a composite of white and pink in the feathers; the deeper flame is the pale crimson of the wings, most apparent in flight. Lesser Flamingos are regularly seen in the company of their larger relatives, the Greater Flamingos, which are taller and of a light pink colour, with a light pink bill instead of a crimson one. Whereas the Lesser Flamingos feed on microscopic lake plants such as algae and diatoms, the Greater Flamingos are carnivorous, consuming larger morsels found in the lakes – crustaceans, molluscs, worms, larvae and small fish.

Lesser Flamingos are rarely seen alone, or in pairs or small numbers; they apparently avoid smaller lakes, even though algae flourish there too. Always watchful for predators, they favour larger lakes, where they congregate in their thousands, because there is safety in numbers. Greater Flamingos, by contrast, are rarely preyed upon and are often found in groups of tens rather than of hundreds or thousands.

A flamingo's wings display distinctive shades of white and pink. The colour is derived from the carotenoid-rich algae the birds feed on almost exclusively.

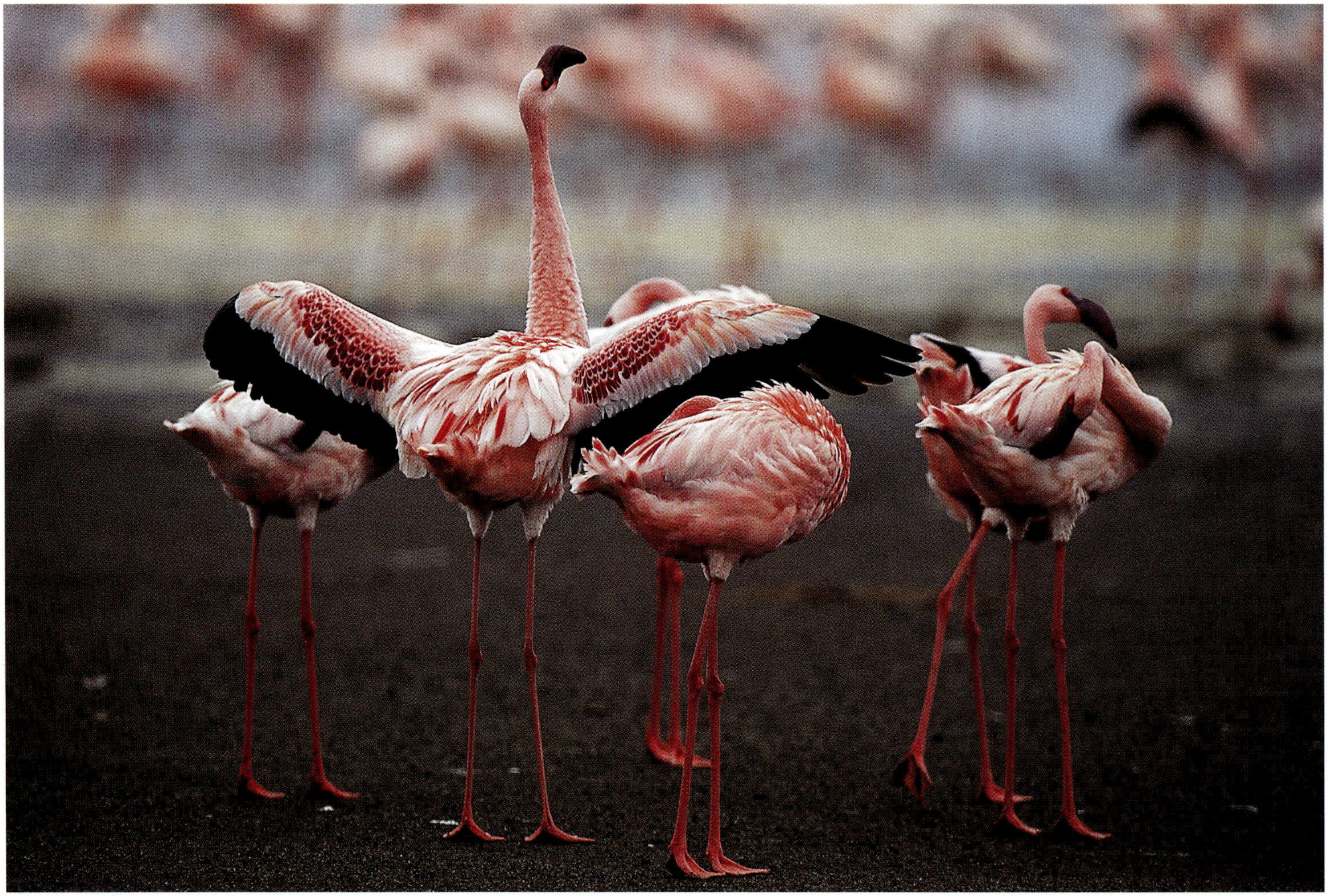

Three key flamingo lakes

Like stepping stones, from north to south, all of the alkaline lakes of the eastern rift are utilised by Lesser Flamingos, but there are three lakes that feature prominently in their life history – lakes Bogoria, Nakuru and Natron.

Lake Bogoria

Lake Bogoria is a strip of blue-green, some 30 square kilometres (11.7 square miles) in area; from the air it looks idyllic, but conditions are harsh on the ground. The shores of Lake Bogoria are hot by day, with some respite in the evenings. The geysers and hot springs on the western shore provide spectacular evidence of ongoing volcanic activity: the molten magma deep under the Earth's crust heats the ground water to extreme temperatures; it is forced under pressure to the surface, where it escapes dramatically through vents into the air. This ceaseless activity is set in a rugged landscape strewn with huge lava boulders.

In this primaeval theatre, the rumble of the flamingos is a constant background hum, as they queue up at the mouth of a steamy, soapy rivulet flowing into the soda lake, carrying with it dissolved carbonates and sodium. In this hot, dry region, the lake, with no outlet and a high level of evaporation, is so alkaline that it cannot support fish, but the concentrated salts are conducive to the growth of algae, and these flourish abundantly, especially near the shoreline. The flamingo feeds here by wading in the shallows, its neck extended downward and its bill floating at the surface; but algae are also found in deeper water where, feeding as it swims, the flamingo bends its neck over in an arch to keep its bill at the surface. In this position it can, from a distance, easily be mistaken for a pink swan.

Feeding almost exclusively on algae, flamingos depend on an extremely narrow – yet abundant – food niche. The birds could almost be said to have 'cornered the market' for algae – but algae crops are fickle: they bloom and they crash. When one crop fails, the flamingos are fortunate in being able to move on to other lakes where the algae are blooming.

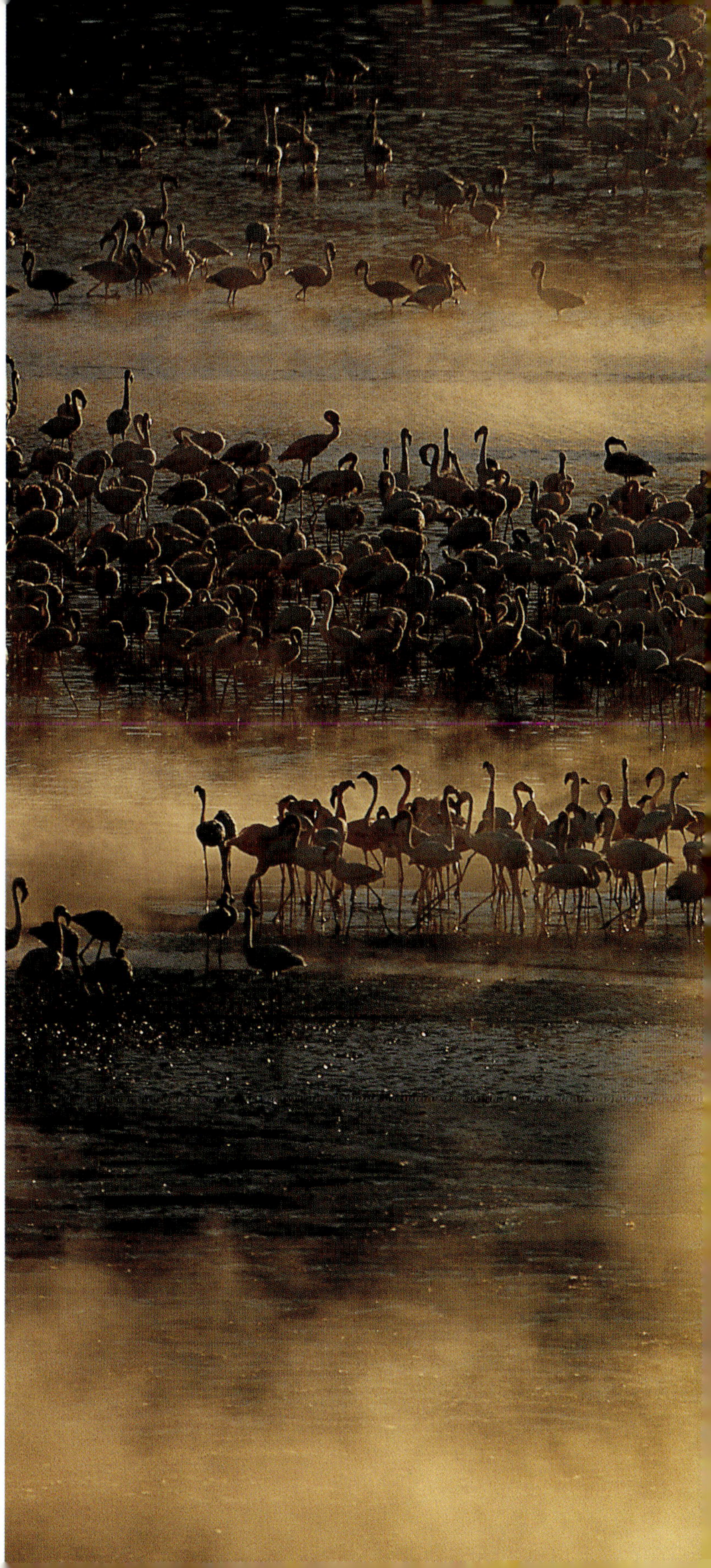

Hot-water springs deliver fresh water to Lake Bogoria. Here, feeding flamingos file past the mouth of a steaming rivulet carrying mineral-rich hot spring water into the lake.

Predators and other dangers

Lake Bogoria is not a breeding lake for flamingos: it offers no haven – neither islands nor mud flats break the lake's interior – on which the birds can safely build nests; on these unprotected shores, both eggs and unguarded chicks would be especially vulnerable to predation. Even juveniles and adults must remain vigilant: on Lake Bogoria, the African Fish Eagle, a consummate hunter of fish, has adapted to hunt flamingos. A flamingo is a heavy item to carry in flight; the eagle solves this problem by hunting and feeding in the shallows. Fish eagles, too, must be vigilant: even warthogs have been known to chase off a fish eagle and devour its kill; for, although a warthog's diet is principally vegetarian, it will seize the opportunity to eat flesh whenever this can be scavenged safely. On rare occasions, a serval, or even a cheetah, has been seen making off with a flamingo in its jaws.

Baboons also launch sporadic attacks on flamingos. One troop, in particular, has hit upon a successful technique: the baboons watch the flamingos bathing at the mouth of a river. Then, as more and more flamingos arrive to join in the bathing frenzy, a weak flamingo or a straggler is spotted; one of the baboons rushes in, trying – sometimes successfully – to grasp the targeted bird with its hands, as they both run splashing through the shallows.

The dangers facing flamingos are not confined to predators, however. In 2000, an unprecedented natural disaster claimed the lives of 50 000 flamingos on Lake Bogoria. The birds were left staggering and convulsing, before dying with their necks curved backward, all showing symptoms of poisoning by cyanotoxins linked to two species of algae in the lake. Post-mortem analysis revealed lethal concentrations of these poisons in the birds' tissues.

Flamingos are continually under threat from such natural traps and ambushes; but feeding, drinking, bathing, preening and sleeping are all parts of life, and the birds cannot be on the lookout for danger all the time. In choosing the strategy of living gregariously in large flocks, flamingos greatly increase their prospects of survival.

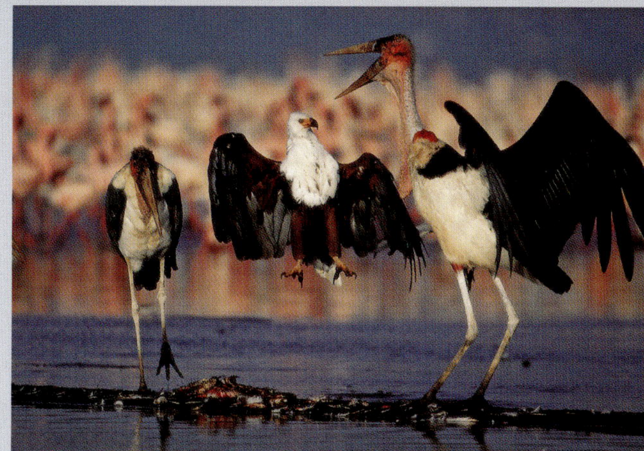

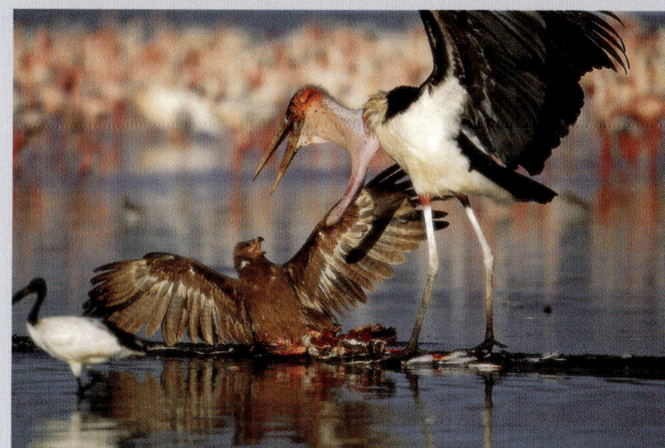

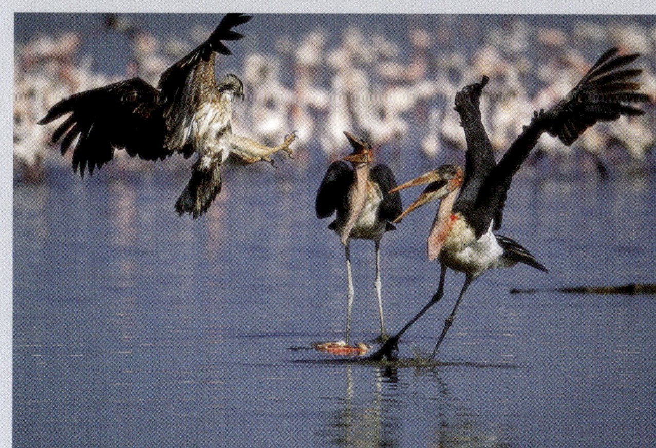

Top: Marabou Storks, Leptoptilos crumeniferus, *arrive to scavenge a share of a carcass; the smaller African Fish Eagle,* Haliaeetus vocifer, *defends its kill.* *Middle: Steppe Eagles,* Aquila nipalensis, *are intimidating scavengers, but no match for the larger Marabou Storks. Right: An immature African Fish Eagle confronts a Marabou Stork over a flamingo carcass.*

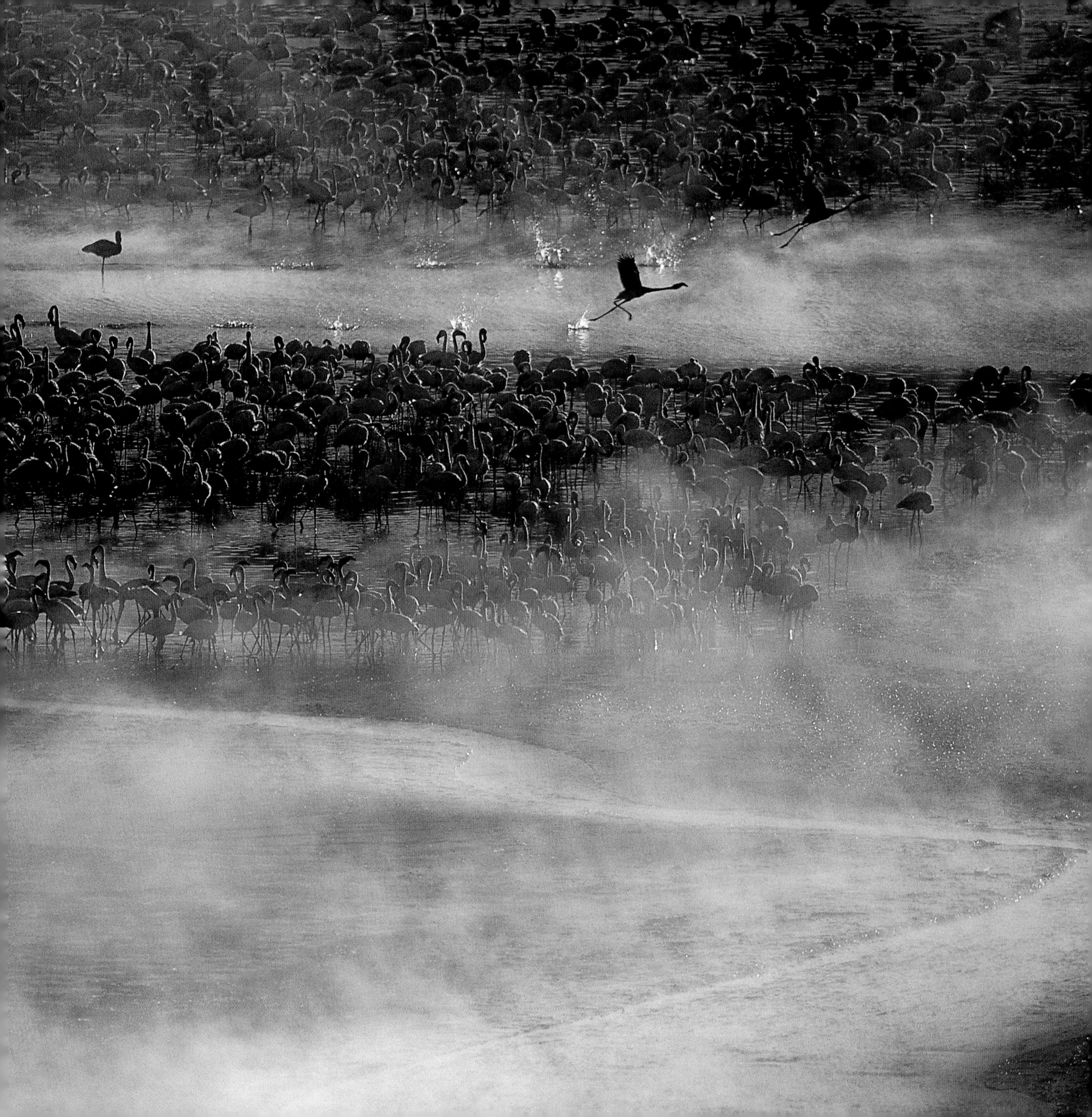

Lake Nakuru

Whereas Lake Bogoria, lying at a low altitude, is hot and arid, Lake Nakuru, being more elevated, is cooler and wetter. Like Bogoria, Lake Nakuru has no outlet, and depends on seasonal rivers and rainfall for replenishment – but water delivery is erratic. The lake itself is very shallow, and so its surface area expands and contracts with the seasons. It is the site for both vast but orderly communal bathing rituals and flamboyantly spectacular pre-breeding displays.

There are a couple of spots on the southern shoreline where flamingos queue up at fresh-water inlets to bathe and to drink. A bathing bird flops down into the water and, with its body partially submerged, flaps its wings vigorously, washing off the salts that stick to its feathers. This lasts only a few minutes. The long-legged flamingo then drinks by bending its long neck down, collecting water in its beak, and lifting its neck up to swallow. A perpetual rumble accompanies these activities, adding to the apparent chaos – yet there is an underlying order. New arrivals join the bathing queue and those that have completed their ablutions join the exit queue.

When bathing, a flamingo sits down and repeatedly splashes itself with water. Flamingos bathe in and drink fresh water.

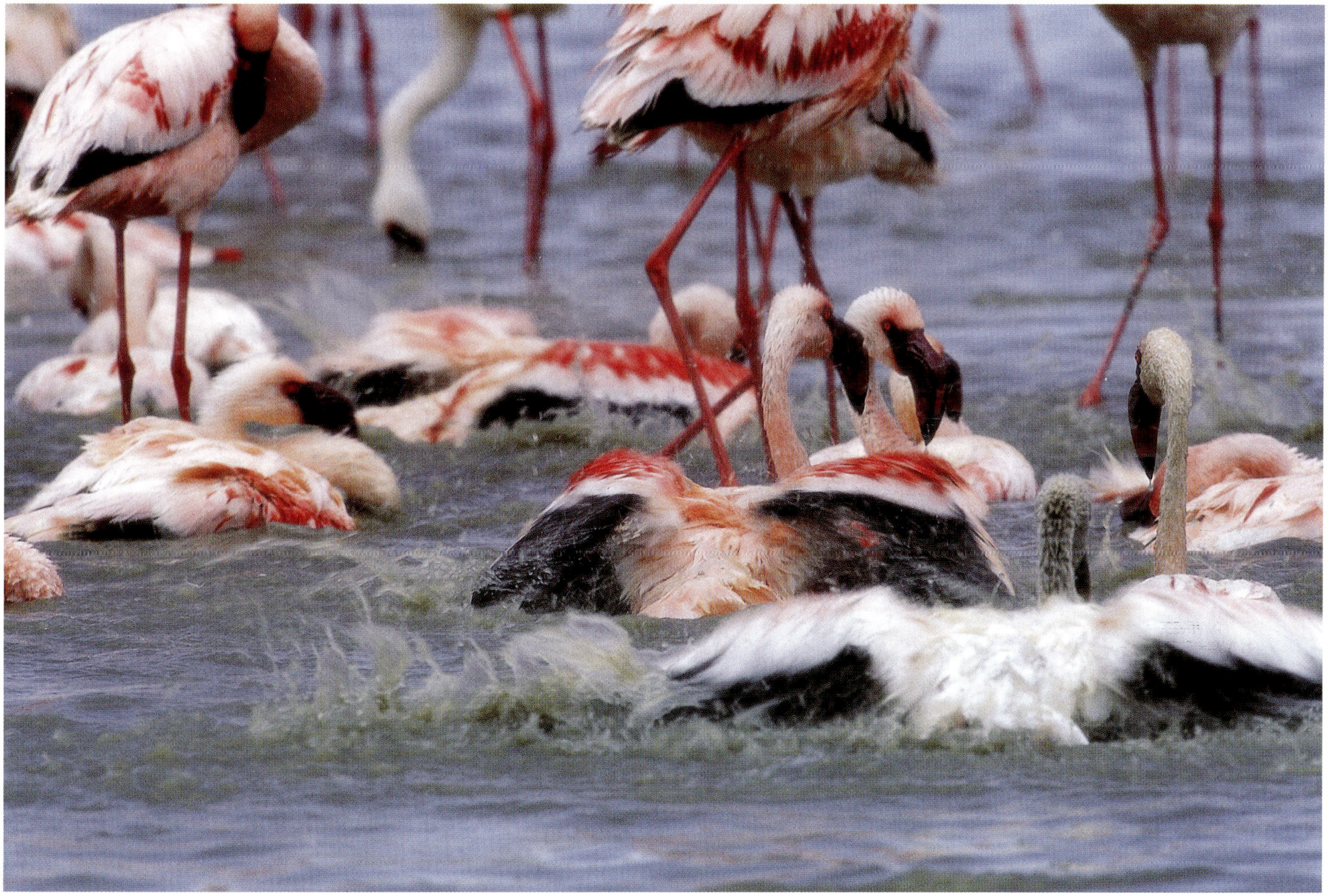

Lake Nakuru's bizarre communal dance

Lesser Flamingos engage in spectacular communal pre-breeding displays, rather like folk-dancing festivals, which happen daily over several weeks. These events tend to occur at established sites, of which there are two on the shores of Lake Nakuru.

A group of birds forms to begin the ritual. Neck feathers partly lifted, these birds raise a tuft of red on each side of their tails as an invitation to surrounding birds to join the communal dance. Soon thousands of participants are packed tightly together; some thrust their heads upward like periscopes which they then flag from side to side, while other birds indulge in 'sword fighting' with their bills, or adopt 'broken neck' postures, sharply kinking their necks to drop their heads and bills down on to their breasts.

The ritual then takes on the semblance of a military parade: a mass of pink-uniformed birds marching swiftly along proceeds abruptly, without visible signal, to change direction with superb synchrony. No bird gets pushed or jostled, or trips up, as the apparently possessed mass of marching birds pushes its way as one body through the main flock, drawing in more birds as it advances, like a huge protean beast surging forward – its multiple necks upstretched, its innumerable legs a twinkling red forest.

On the fringes of the mass procession, individual birds perform wing salutes, twist-preens and wing-flash bows. These exquisitely choreographed performances are orchestrated throughout with a sustained chorus of sound – loud calls, low grunts and, in the background, a subdued roar. Were it not for the synchrony of their movements, the flamingos would appear to have lost their senses.

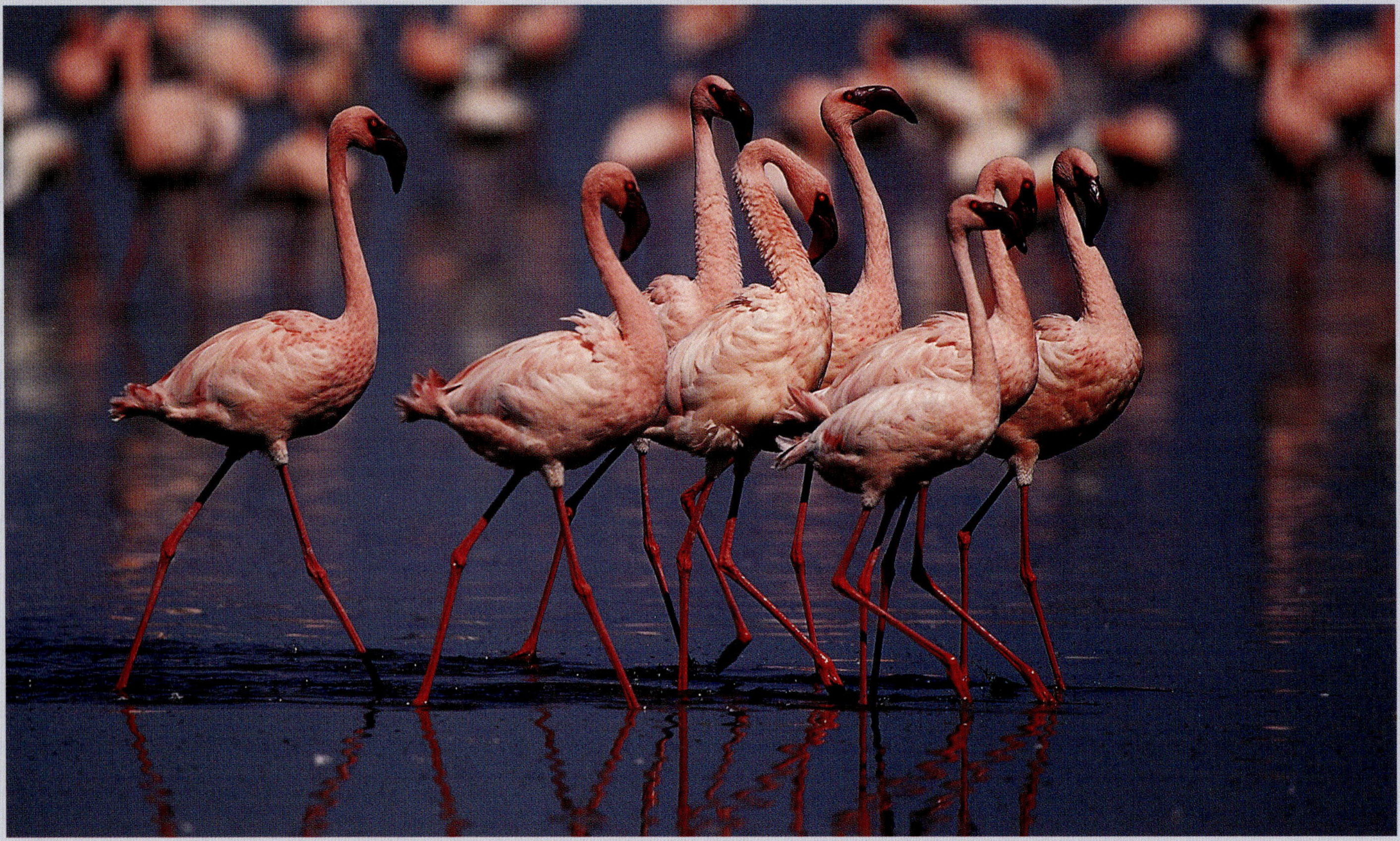

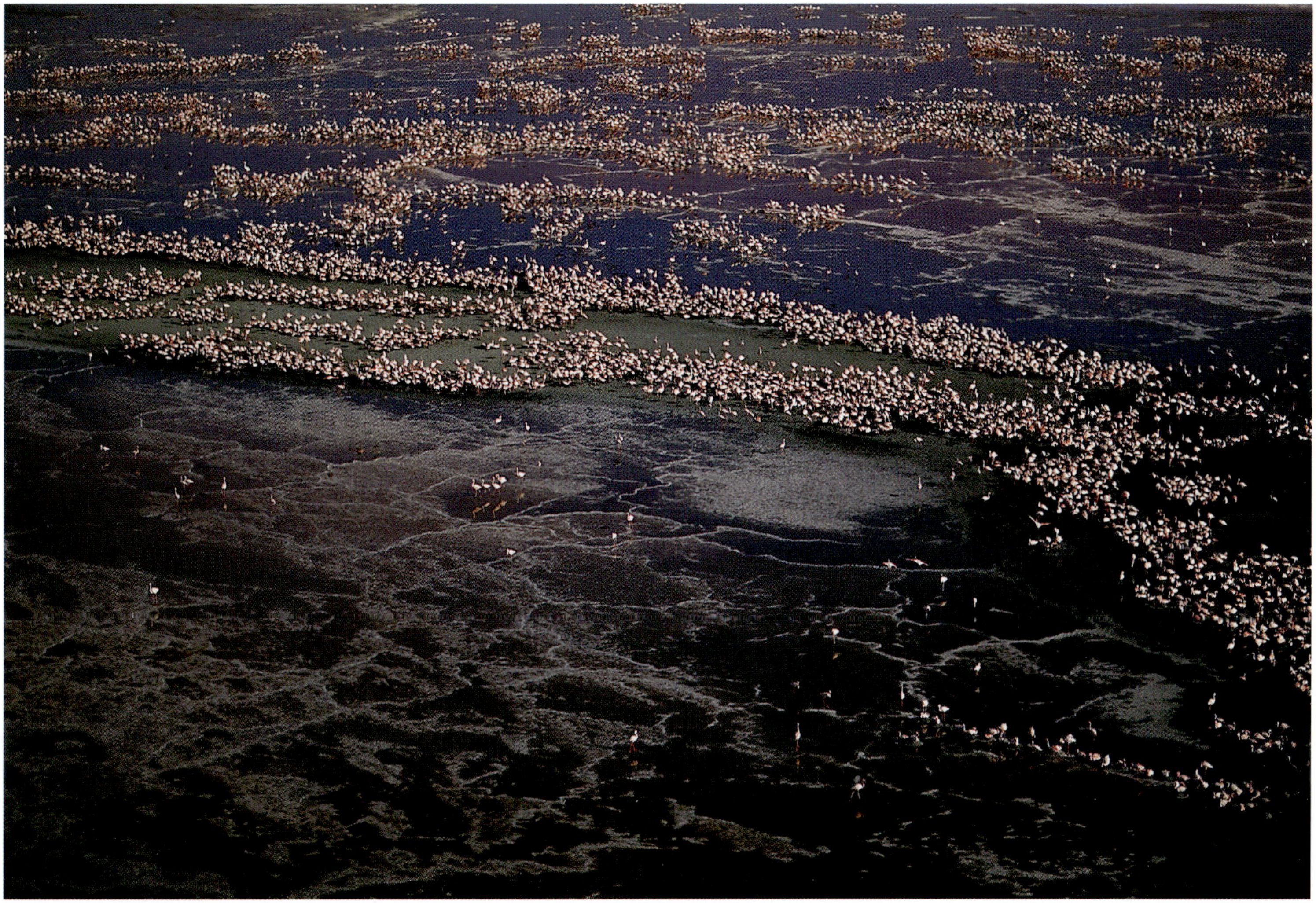

Lake Natron

The communal dances on Lake Nakuru synchronise the flamingos in a heightened state; but neither Lake Nakuru nor Lake Bogoria is safe for breeding. To complete the crucial next step in their life cycle, the flamingos must move to another lake.

Remote, pristine, mysterious and uninviting, Lake Natron lies in northern Tanzania. Although much larger than Lake Nakuru, it is also shallow, and fills up with water for only a few weeks each year when the rains drain into the lake from the Kenyan highlands to the north. That is when it is frequented by flamingos; for the rest of the year Lake Natron is largely dry, exposing a brittle crust of sodium bicarbonate covering vast expanses of soft, slimy mud. Even when the lake has water, conditions are extremely hot and dry, and a foul smell hangs in the air. Relatively few flamingos line the shores of this lake. Some gather at the mouths of the two rivers that flow into the lake; occasionally a group appears in mid-lake; but most of the flamingos are on the mud flats, a long way from the shore, engaged in the crucial activity that makes this the most important of all the soda lakes – for it is the site of the flamingo breeding grounds.

Lake Natron is a closed-basin, alkaline lake disposed in an inhospitable, low-lying region of northern Tanzania. It is one of the few sites in Africa where flamingos breed regularly.

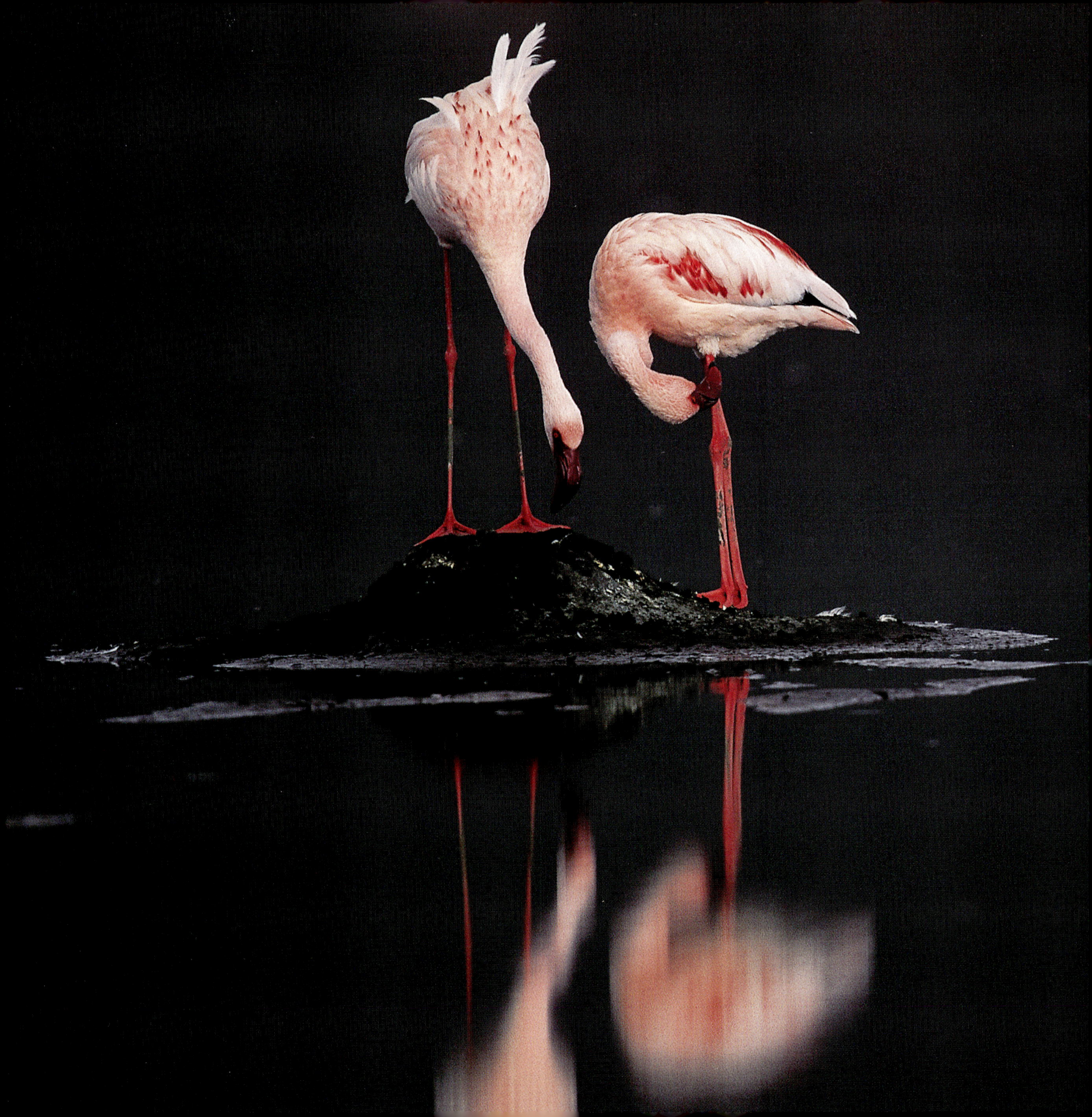

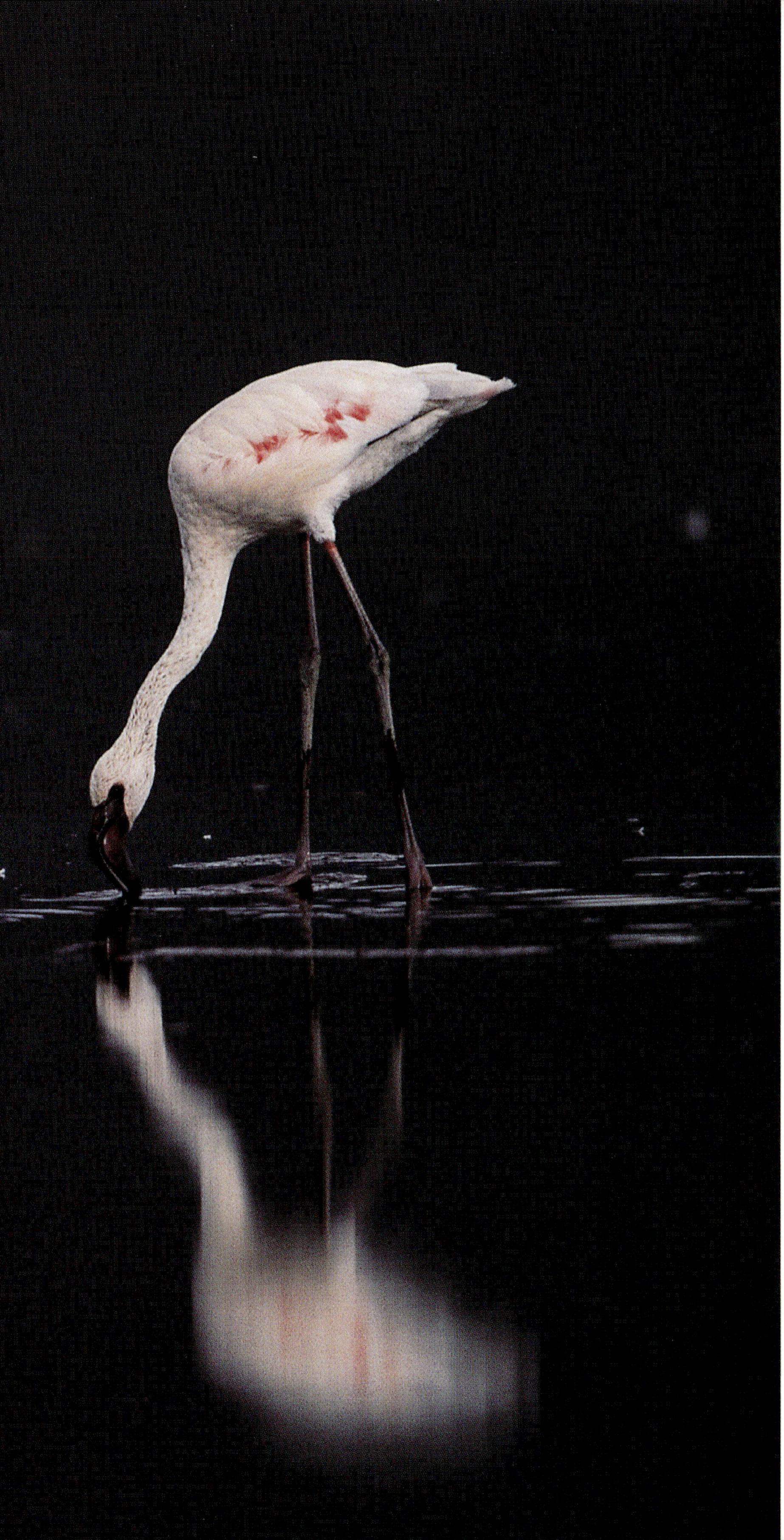

Parent birds scrape mud from the lake floor to build conical-shaped nests on the mud flats in the middle of Lake Natron. The nesting area provides a safe haven for hatching and rearing chicks, out of reach of land-based predators.

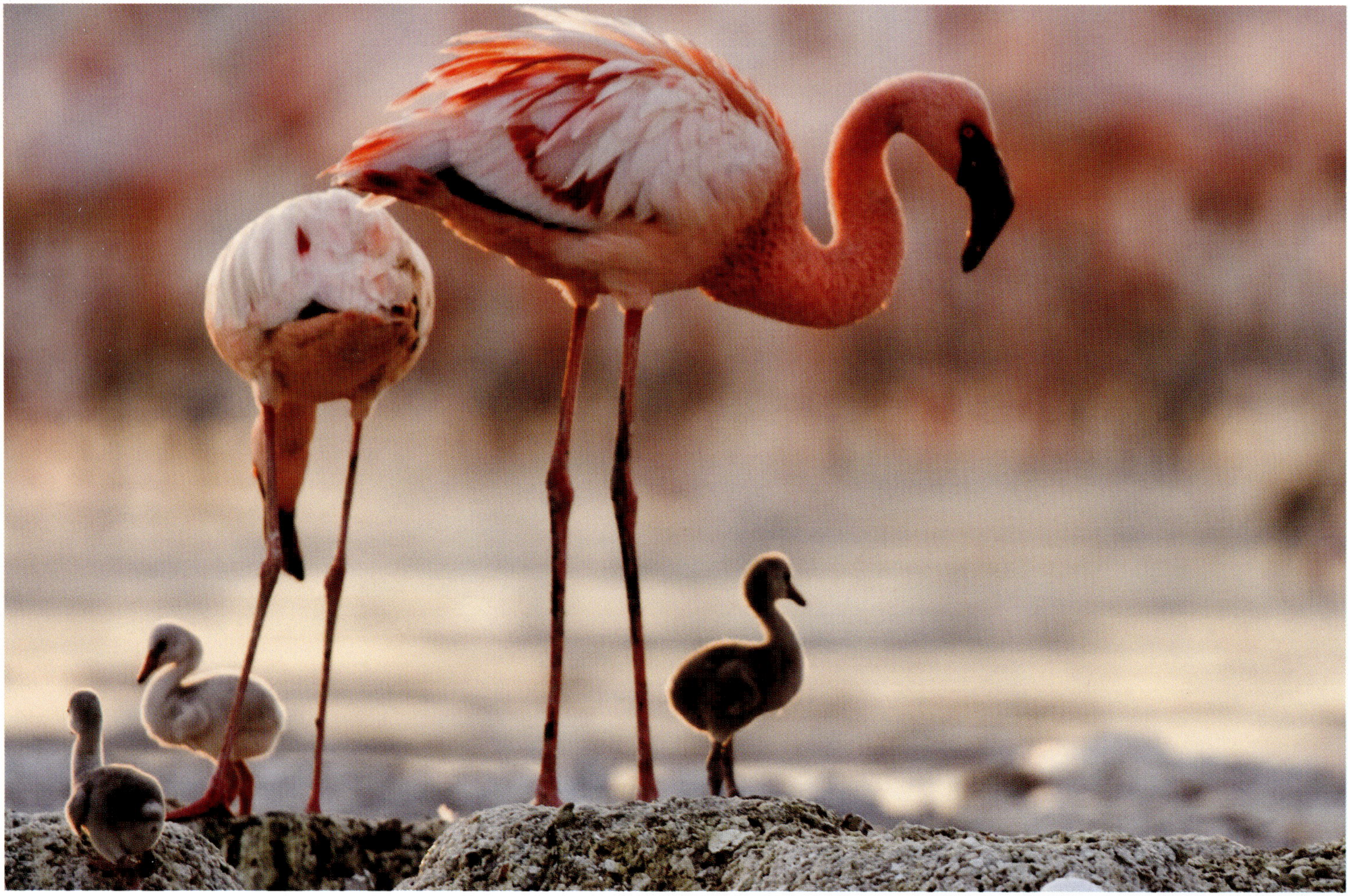

Adults and their young on adjacent nests in the soda-encrusted mud flats in the middle of Lake Natron. The female lays a single egg.

Breeding

Flamingos become sexually mature at around six to seven years, when they normally pair for life. The female lays a single egg, and both parents care for the juvenile for several years. This pattern is repeated until death, at around 40 years of age. The chicks are protected from predation; Lake Natron, with exposed mud flats at its centre, out of the reach of land predators when the lake has water, provides them with a safe haven. Even avian predators are nomadic opportunists rather than resident hunters, since they cannot rely on the timing of flamingo breeding events. When the flamingos arrive, they have the mud flats virtually to themselves.

Flamingos breed en masse. It is difficult to predict the timing of such an event, as there is no regular pattern to their breeding and, inexplicably, only around one fifth of adults participate in any one breeding attempt. When the time is right for the flamingos to breed, they form a huge colony in this large and shallow lake. Turret-like nests are constructed from mud along junctions in the soda plates where the mud is softer and more pliable. The built nest is about 30 centimetres (one foot) high, and broad enough for a parent bird to sit on. The finished colony is a crazy mosaic of short, irregular lines of nests, topped by birds with writhing necks and wide-open bills – each defending its claimed territory.

The largest, hottest soda lake in the rift valley, with an alkalinity of up to 11 pH, Lake Natron is surely one of the harshest nurseries in the world. The temperature of its surface mud has been measured at 60–65 °C (140–150 °F). Without shade, humans could not survive the harsh conditions to which the flamingos are so well adapted.

Two weeks after hatching, the chicks undergo an arduous ordeal; strung out in lines as much as 30 metres (100 feet) long, they must trudge across the mud flats toward permanent, freshwater lagoons. Conditions on Lake Natron are highly unpredictable. An abnormally long, hot, dry spell sometimes prevails, causing an increase in the concentration of soda; under such conditions, anklets of soda can accumulate around the legs of the flamingo chicks. By 10 weeks they can fly, but until then they have to walk, even on legs encrusted with dried salt. Some will stumble and fall, helpless to right themselves, and die a painful, lingering death. At the other extreme, there is the danger of unexpected flooding. When the rising water level submerges the nests, the adults cut their losses and abandon their eggs or hatched chicks – a harsh reminder that here, Africa is still untamed.

Early development and life strategy

In the absence of droughts or floods, growing up is a smooth process. The chicks, fed on bright red, protein-rich crop milk, grow fast. Eight days after hatching, they are able to leave the nest and swim. At around two weeks, they are grouped into crèches, and 10 weeks later, they are weaned off crop milk and encouraged to feed for themselves. By this time, their bills have developed and they have acquired the full set of feathers necessary for flight. Despite this relatively short period of direct dependence on their parents, flamingos reach sexual maturity only after six or seven years.

This brief period of infancy and a drawn-out period of immaturity is followed by a very long adulthood. Those offspring that survive to sexual maturity join the huge flocks of flamingos that will embark on subsequent breeding attempts. Their successes will ensure that the flamingos' reproductive gamble has paid off handsomely.

A young flamingo falls prey to a spotted hyaena, Crocuta crocuta, *at Lake Nakuru National Park in Kenya.*

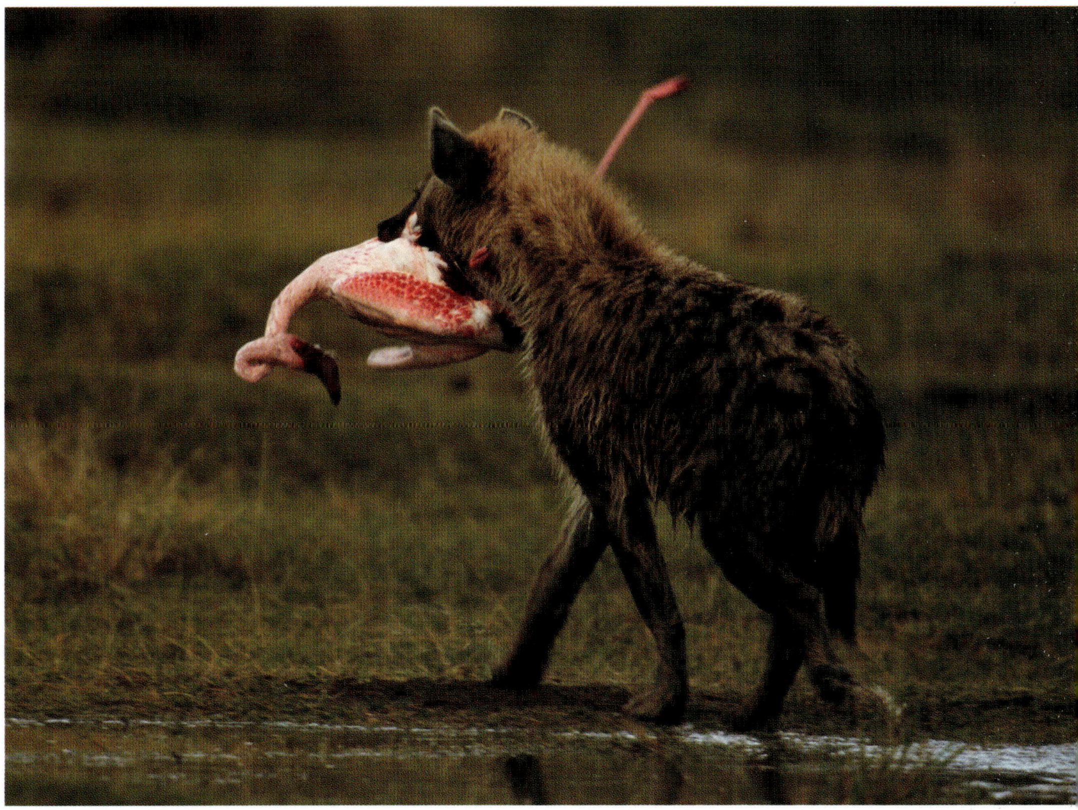

Exploiting a unique niche

There is nothing to temper the relationship between flamingos and their harsh environment. From birth until death, they are exposed to the unremittingly bleak and hostile conditions of the soda lakes, and yet they survive, reproduce and currently number four to five million – a triumph of adaptation. In the primitive environment of the soda lakes of the Great Rift Valley, this ancient life form exploits a unique and stable niche, honed through evolution by natural selection over millions of years.

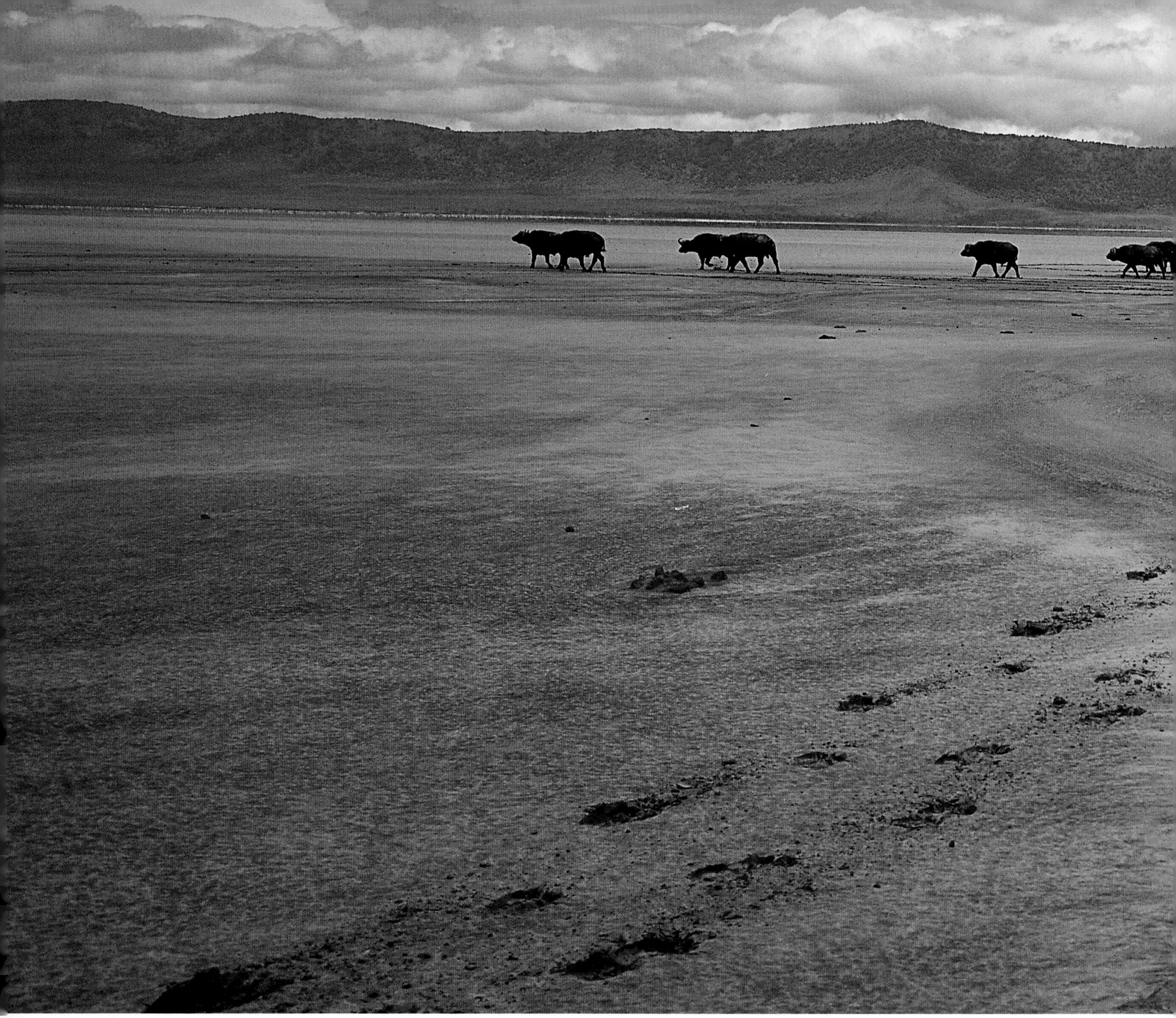

A small herd of Cape buffalo bulls, Syncerus caffer, *leaves behind a trail of footprints on the exposed mud of a dried-up area of Lake Makat, inside the Ngorongoro Crater.*

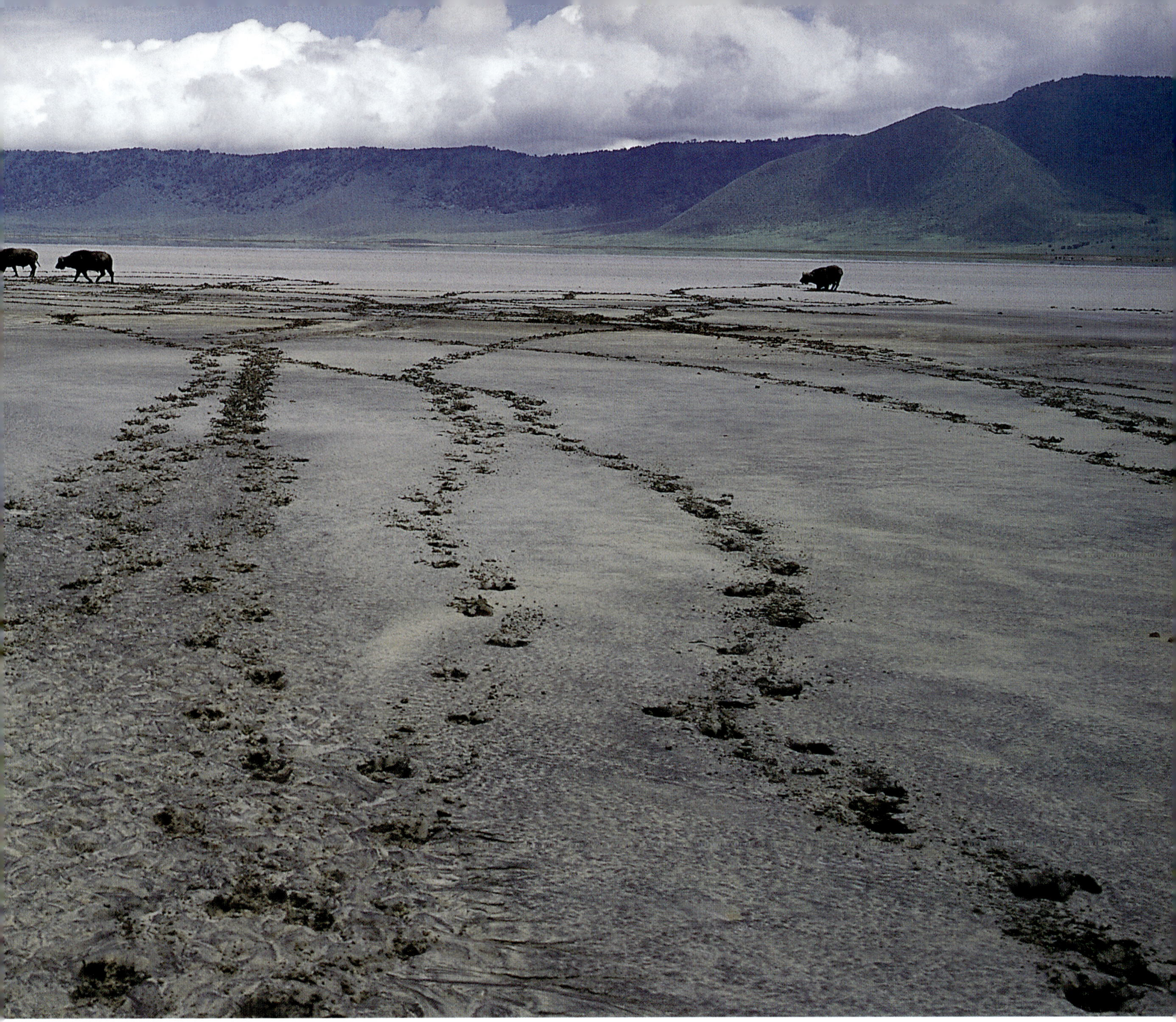

3 Wildlife of the Crater Highlands

About five million years ago, as their habitat shrank, some of the forest-dwelling apes moved east to the adjacent wooded savannah, where they lived for the next 1.5 million years. One day, two adults and an adolescent walked across a small mud flat; simultaneously, a nearby volcano started erupting, and covered their footprints with a layer of ash. Light rain fell; the sun came out, and hardened the ash. Thus, by a serendipitous combination of events, these footprints were preserved, to be uncovered in 1978 – 3.6 million years later – as the earliest known evidence of an ape that walked on two feet. Baboons, gorillas and chimpanzees use their forepaws, or hands, as feet. But these individuals, subsequently named 'southern apes', were different: their hands no longer fulfilled any locomotive function; they stood and walked upright. The 'southern ape' would have resembled a modern chimpanzee, distinguished, however, by its posture, the shape of its hips, and its hands, which were free to manipulate objects by grasping them, picking them up and carrying them. It could also travel much further and for longer periods than any of its primate predecessors.

Below: Footprints found at Laetoli in the Crater Highlands, identified as being 3.6 million years old, are the earliest evidence of bipedalism in the ancestors of humankind. Right: An aerial view of a shoulder of Lemagrat Volcano shows deeply cut gullies, resulting from lava flows in earlier eruptions.

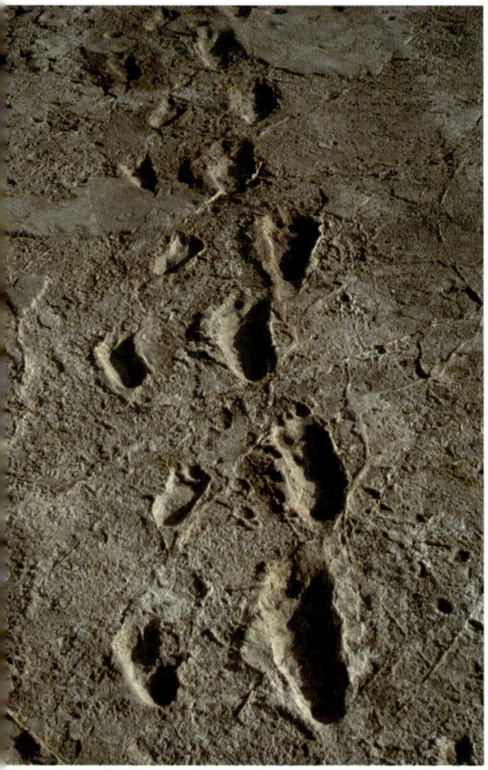

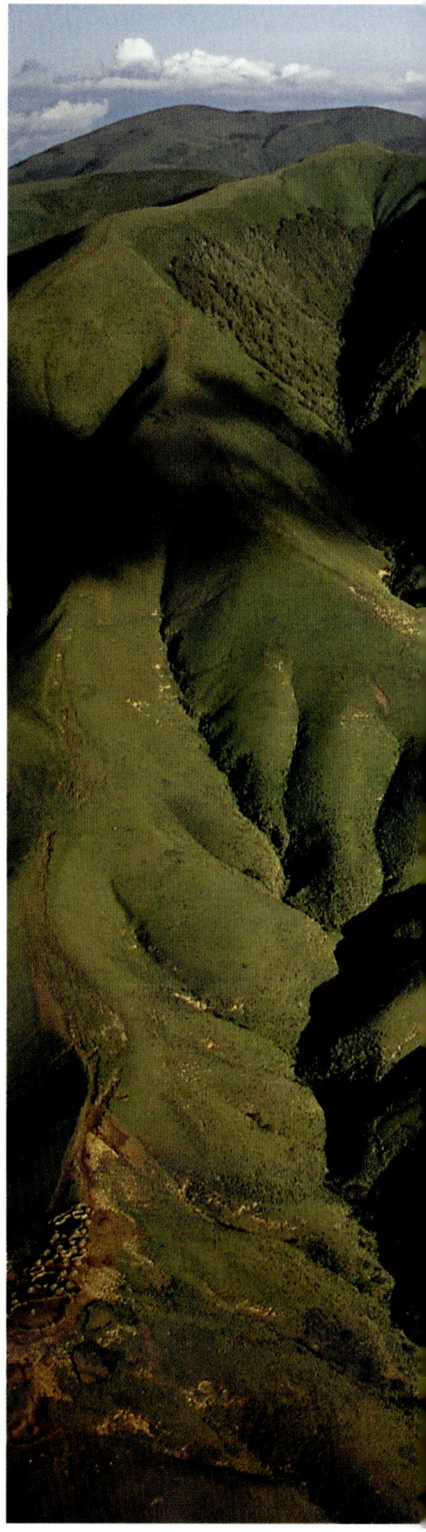

The formation of the Crater Highlands

These footprints were found at Laetoli, lying at the base of the Crater Highlands of Tanzania, where the superb setting of towering mountains, volcanoes, craters, escarpments, forests and plains reflects the impact of the continental rift on an ancient landscape. The onset of the rifting, about 20 million years ago, thrust up a long, winding cliff as the land sank. Most of this early rift is now obscured by lava that, having welled up in the fracture, began overflowing some five million years later. The lava built up layer by layer; eventually, the outflow of fresh lava was restricted to a limited number of openings. These built up over millions of years into the volcanoes that are such a prominent feature of this area today. Each volcano has its own distinct layer of basalt and ash, from which geologists can estimate its age. Oldeani, Lemagrut and Sadiman are the oldest volcanoes, and it is Sadiman's eruption, 3.6 million years ago, that discharged the ash that preserved the Laetoli footprints.

Meanwhile, to the east, Ngorongoro, at that time still an active volcano, was slowly accumulating bulk and, 2.5 million years ago, was probably as tall as Kilimanjaro. As its vents filled up with solid rock, the mountain's trapped lava found another outlet, resulting in what must have been a massive eruption. With its underlying support now vented, the summit of Ngorongoro collapsed inward, forming the huge basin of an extremely wide crater, known as a caldera – the largest in this volcano-dense area. The unbroken rim of this caldera can be identified by astronauts in space.

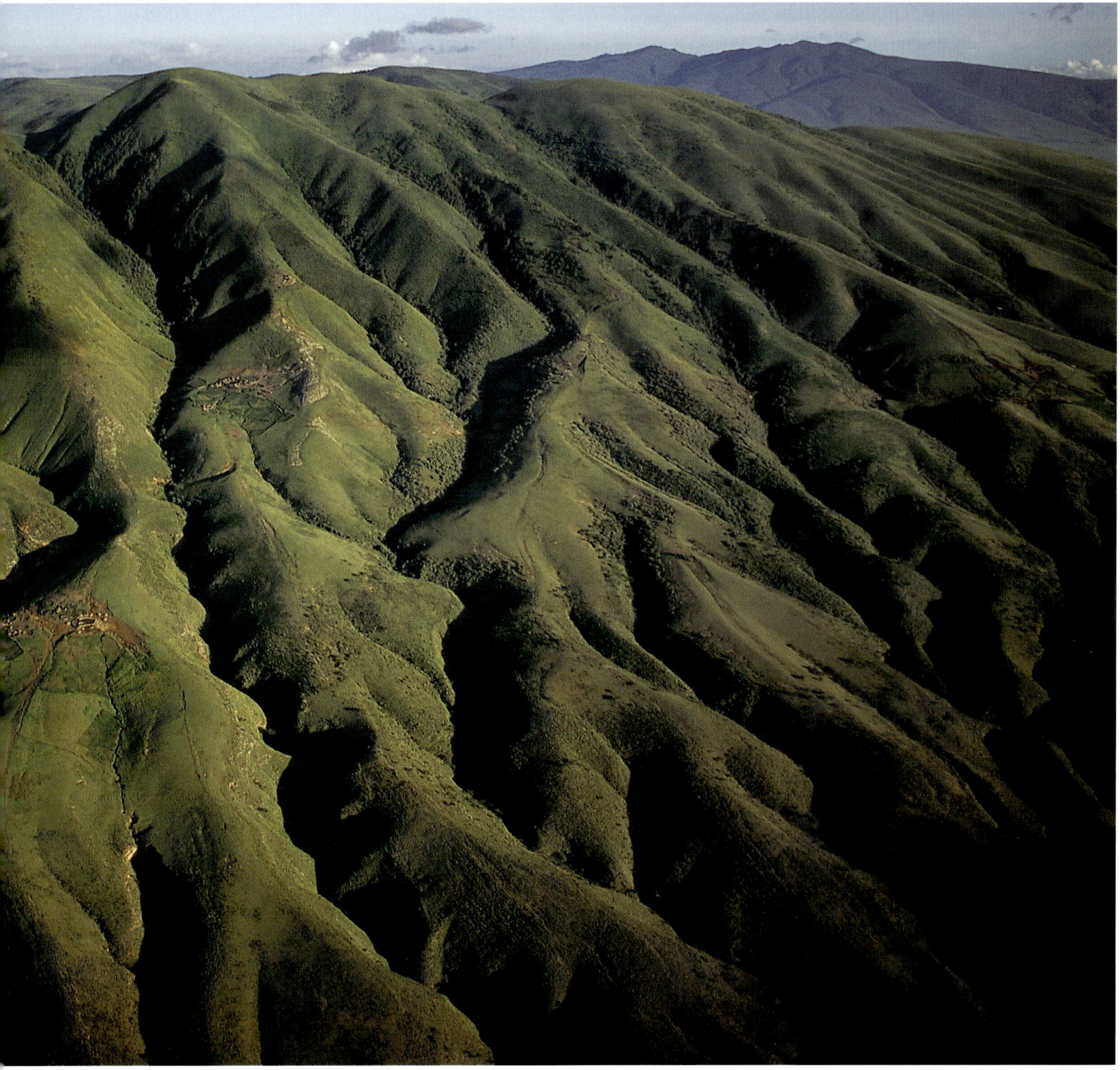

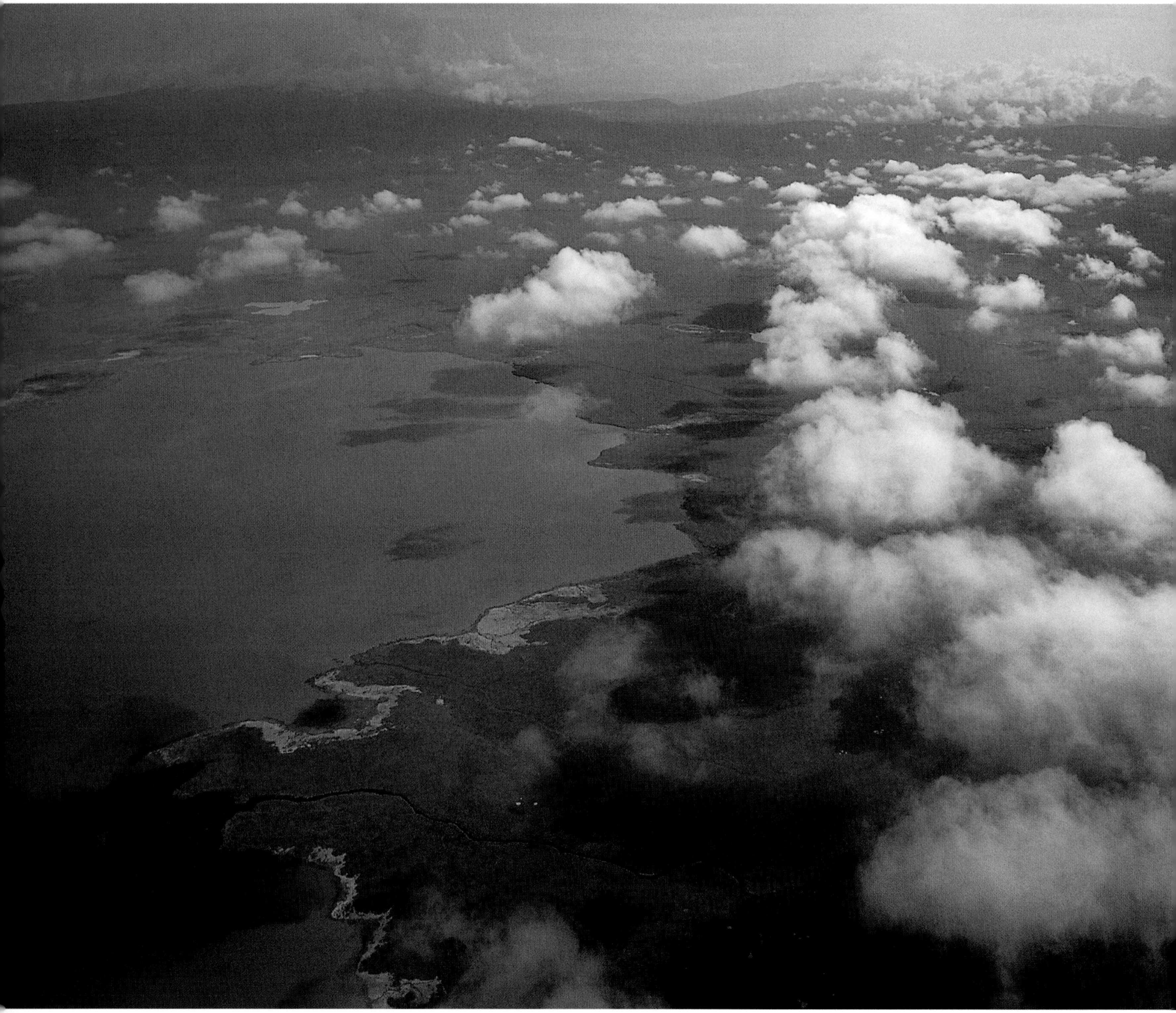

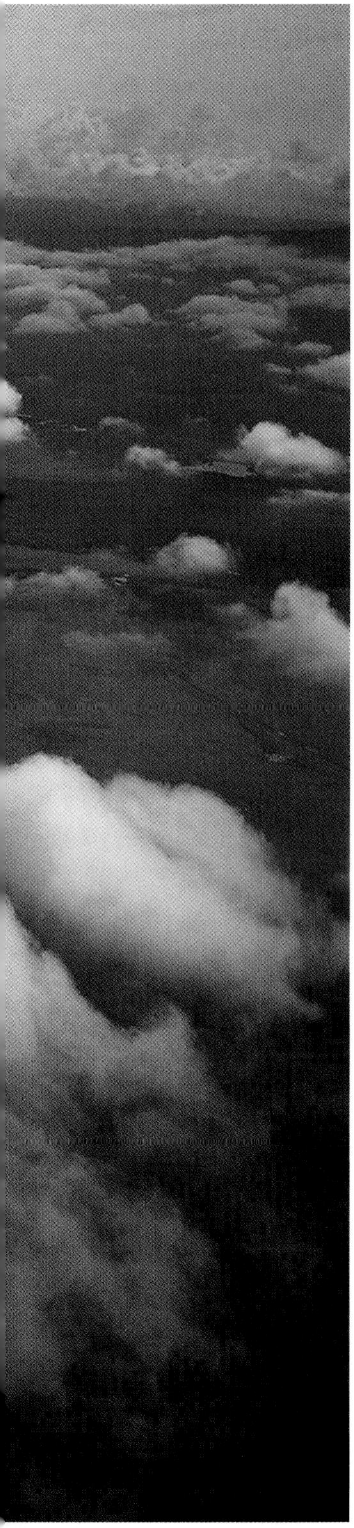

In the same way that Ngorongoro collapsed and developed a caldera, so, too, did Olmoti and Empakaai, to the northeast. These collapses were followed by the rise, further northeast, of the volcano responsible for Serengeti's short grass plains – Kerimasi. As Kerimasi died, Lengai was born to its north, and it is still active today. It bubbles and belches steam from vents in its summit and occasionally puffs out ash. It last erupted in August, 2007.

By two million years ago, the main features of this area had been formed into the landscape we know today. Erosion and rainfall then fine-tuned the sculpting of the land.

Ngorongoro

Crater water

Today, the Crater Highlands are a rash of massive volcanoes and calderas, creating their own climate. This is clearly demonstrated at Ngorongoro Crater, where the clouds that roll up against the crater rim during the wet season shed a lot of rain, much of which is retained by the montane forest surrounding the rim; slowly, throughout the year, some of this moisture is released into the crater. Even during the long dry season from June to October, the mist drifting over the rim adds moisture to the montane forest.

The crater has additional sources of water: the Munge River pours down the northern wall of the crater from its source in the broken caldera of Olmoti Volcano, and springs emerge from the crater wall itself to form swamps on the crater floor. It is not surprising that, soon after the collapse of Ngorongoro's cone, vegetation recolonised the crater, taking advantage of the rich soils and the abundant moisture.

The herbivores followed the vegetation. Over the steep mountain trails, one species after another from the surrounding ecosystems climbed the crater walls and descended into the interior. Inevitably, the carnivores followed the herbivores.

Today, Ngorongoro Crater is like a lost world, maintaining the atmosphere of ancient Africa (if you ignore the tourist traffic). From the circumference rim running for 64 kilometres (40 miles) – unbroken, though indented by watercourses – the crater floor, covering an area of 325 square kilometres (126.8 square miles), lies some 500 metres (1 640 feet) below. It is not of uniform height; the occasional plateaus and rock fragments serve as a gentle reminder of its violent creation. Spread out on the crater floor are many of the features that make the crater a microcosm of the East African plateau: long grass plains, short grass plains, freshwater swamps, a soda lake, a forest, streams and hills. It is this abundant variety of habitats that allows so many species of animal to live together in such a small area. Furthermore, since the crater floor has a year-round supply of water and rich pastures, animals live here in great abundance.

Left: *Rain clouds gather over Ngorongoro Crater. A permanent supply of fresh water is the caldera's great attraction for wildlife.*

Crater floor wildlife

A big herd of the resident population of wildebeest, Connochaetes taurinus, *inside the Ngorongoro Crater.*

There are between 20 000 and 25 000 large mammals in the crater. They are not confined by the crater walls, even though the rim of the crater is steeper on the inner than the outer face – this is the normal form of volcanoes. There are places where the animals can choose to walk up and over the natural boundary. They stay because the incentive to leave is weak.

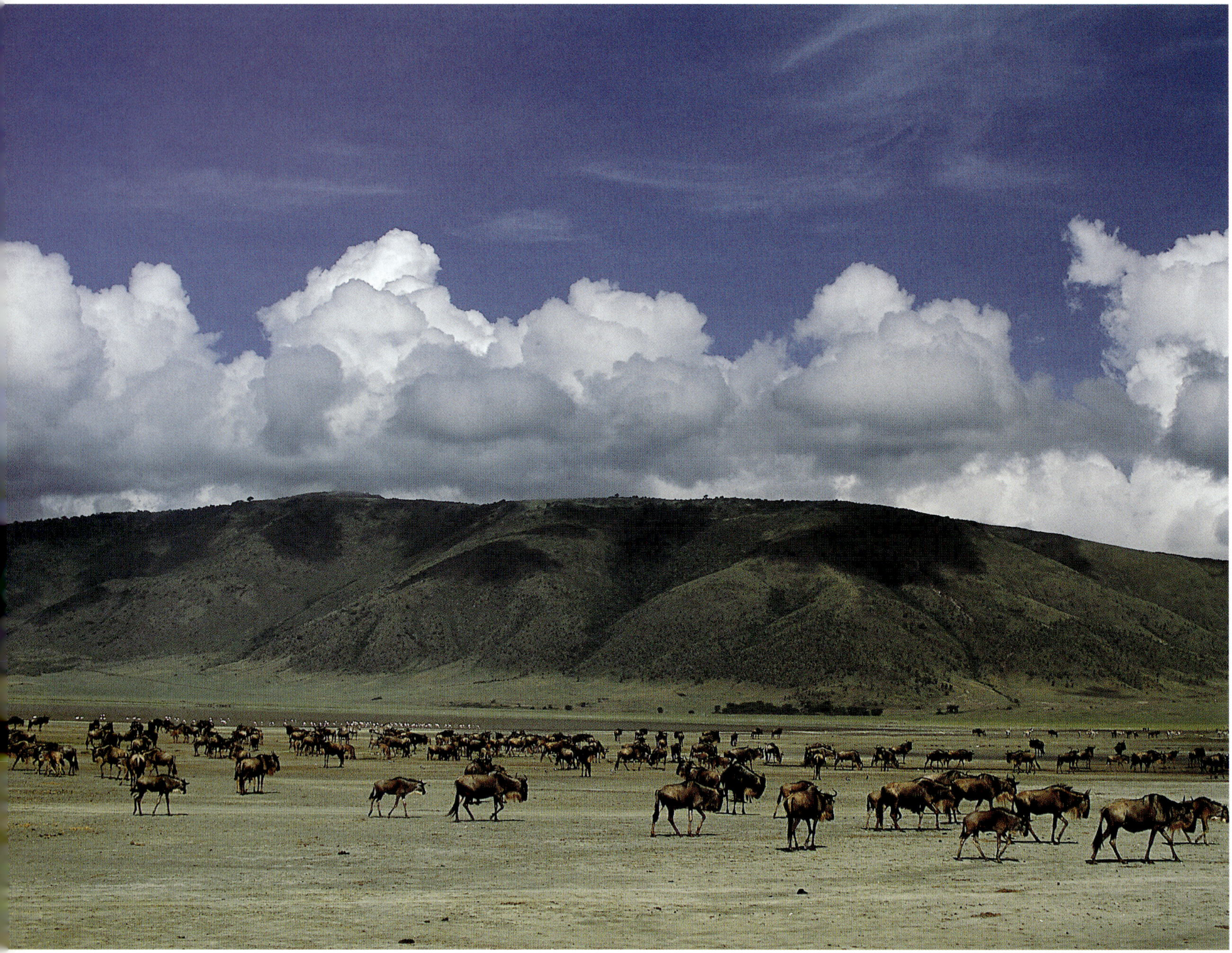

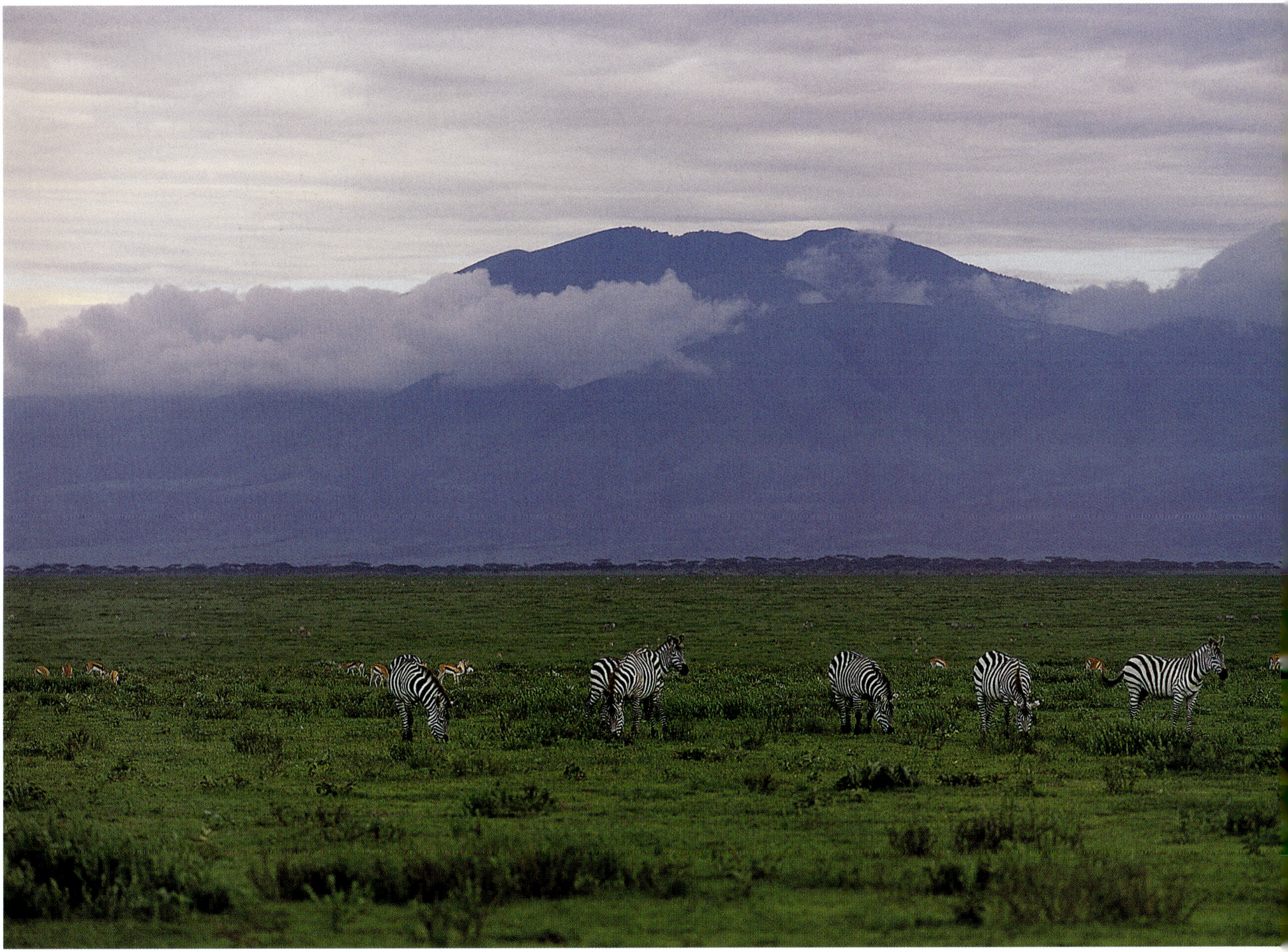

Nevertheless, the crater is a dynamic and changing ecosystem. Since the mid-1950s, censuses have shown that wildebeest numbers have fluctuated between 8 000 and 15 000. Both warthog and buffalo first appeared in around 1976. Black rhino numbers have decreased from 100 in 1965 to fewer than 20 today, due mainly to poaching. Some species, among them vervet monkeys, have decreased in number, losing their habitat as the Lerai Forest in the crater thinned out. Hippos, though, have maintained a stable population, limited by available freshwater resting spots.

Migratory zebra, Equus burchelli, *and Thomson's gazelle,* Gazella thomsoni, *graze on newly sprouted green grass near Olduvai Gorge. In the dry season the grazers move on.*

Right: *An aerial view of Maasai giraffe,* Giraffa camelopardalis tippelskirchi, *on the western rim of Ngorongoro Crater, where they are frequently encountered near the acacia trees on which they browse. Although conspicuous in the region, giraffe are not found inside the crater.*

Below: *Lerai Forest inside Ngorongoro Crater is populated by several troops of acrobatic vervet monkeys,* Cercopithecus aethiops.

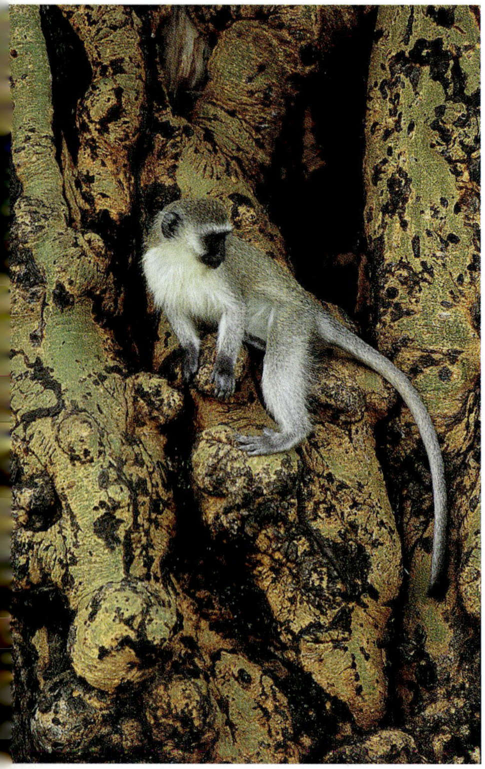

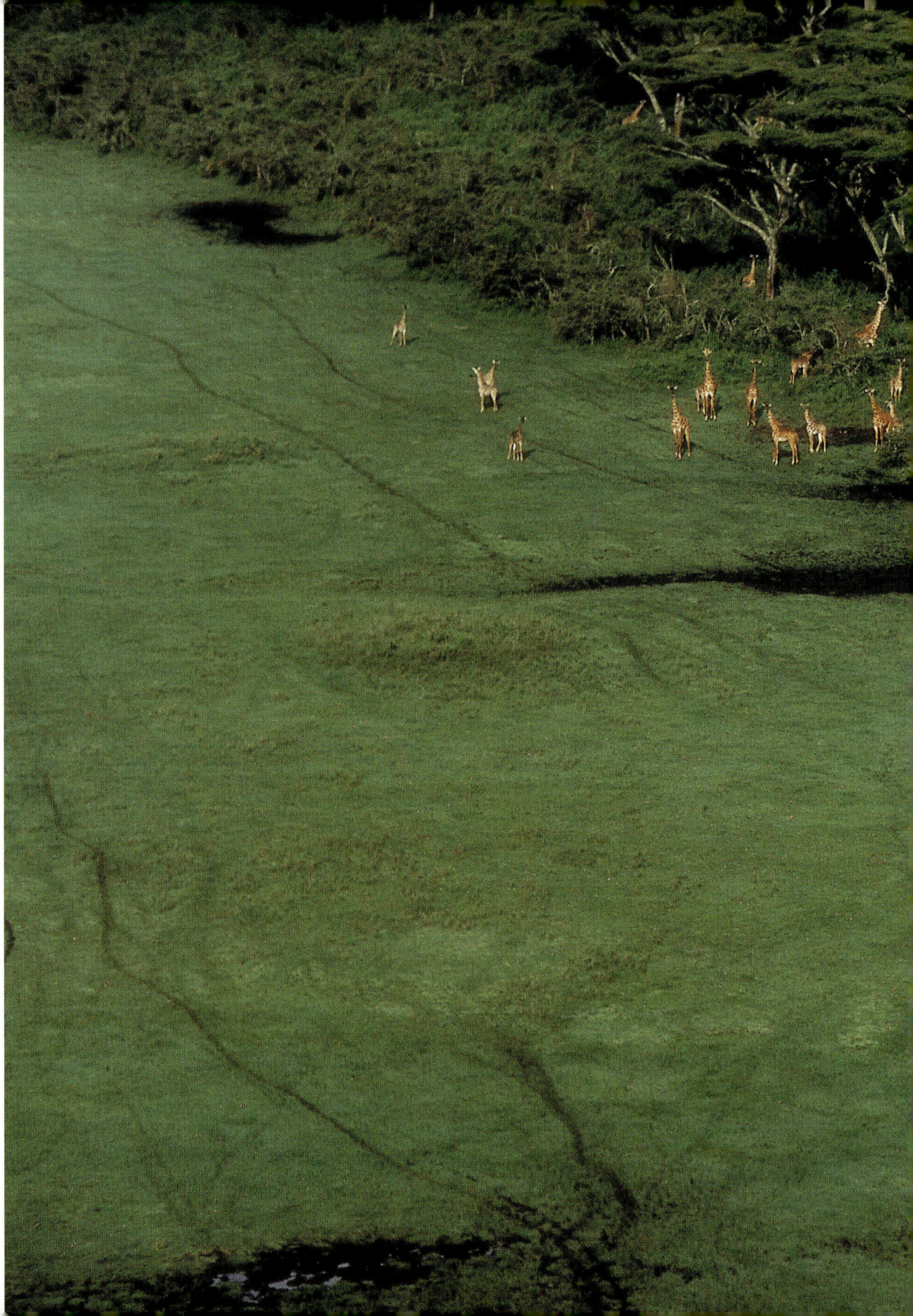

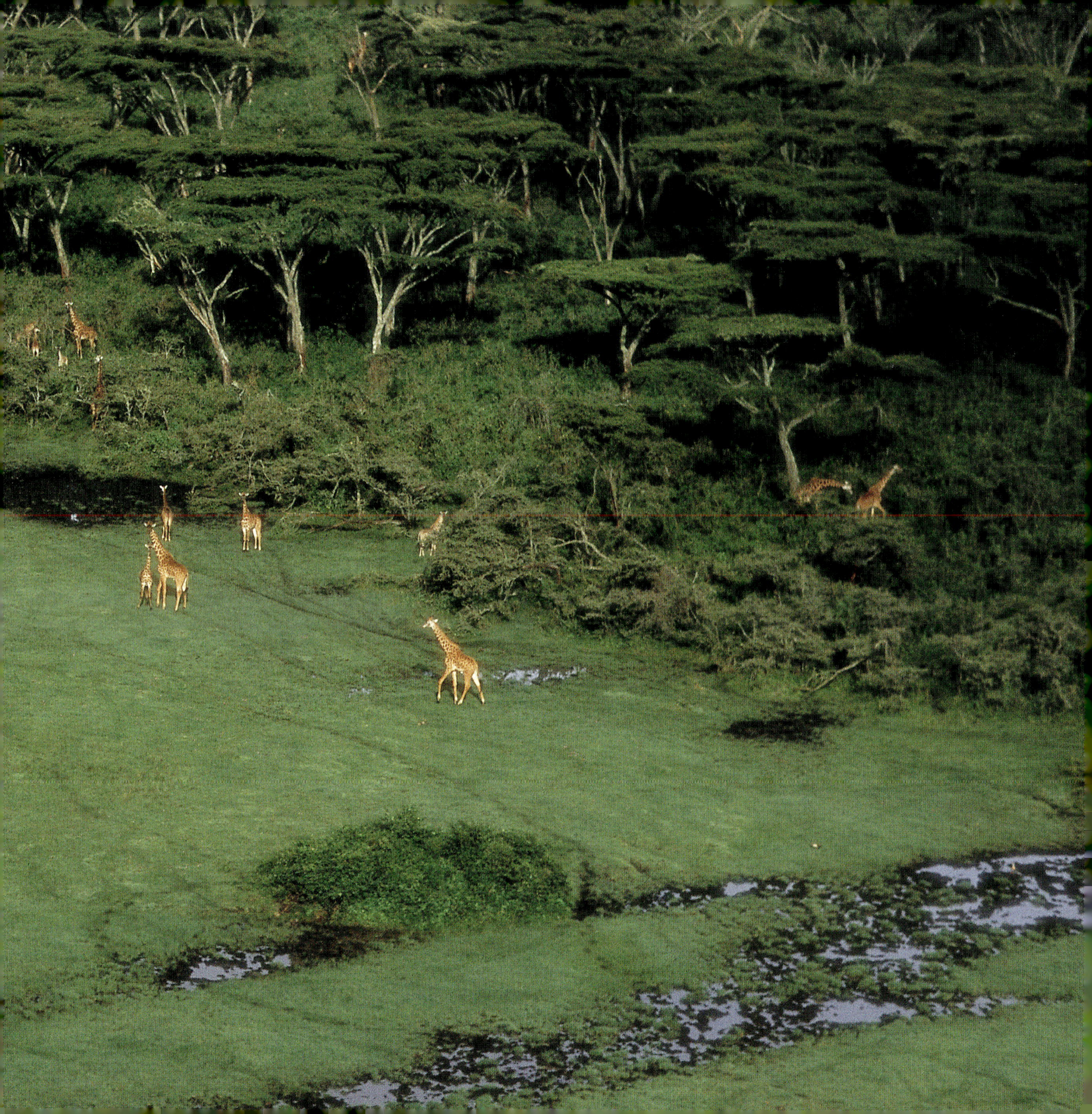

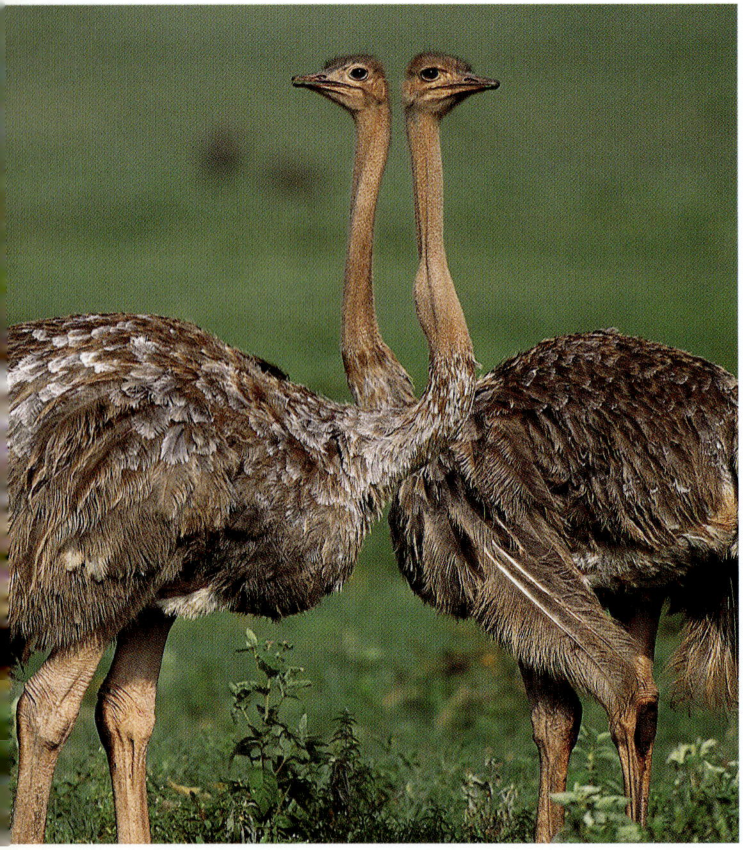

With so many prey animals, carnivores are abundant. Lion numbers have fluctuated, with an upper limit of 100 individuals; spotted hyaenas have reached four times this number. There are five main lion prides and eight hyaena clans living on the crater floor. Such dense populations of large predators are unmatched elsewhere in the world.

The crater floor is also home to the large birds of Africa: migratory crowned cranes, nomadic flamingos, and resident ostriches and Kori Bustards. Lake Makat on the crater floor attracts a multitude of birds; the variety of species present at any one time is determined by the dynamically changing mixture of soda and fresh water in the lake. Under certain conditions, there is a bloom of algae, turning the lake water a pea-green colour. It is not long, then, before the lake edges turn pink, laced with flamingos. Opportunistic golden jackals will in turn take up residence around the lake to hunt the flamingos.

Left: Ostriches, Struthio camelus, *are well adapted to life on the open plains inside Ngorongoro Crater.*

Below left: Hippos, Hippopotamus amphibius, *courting in a hippo pool in the crater.*

Below: A newborn wildebeest calf in Ngorongoro Crater staggers towards a wildebeest female that unhesitatingly lowers her head to steer it away. No wildebeest cow will allow a calf that is not her own near her.

Opposite: Two young zebra stallions engage in a bout of play-fighting that helps to establish the dominance order within a bachelor group. The zebras remain in the crater throughout the year.

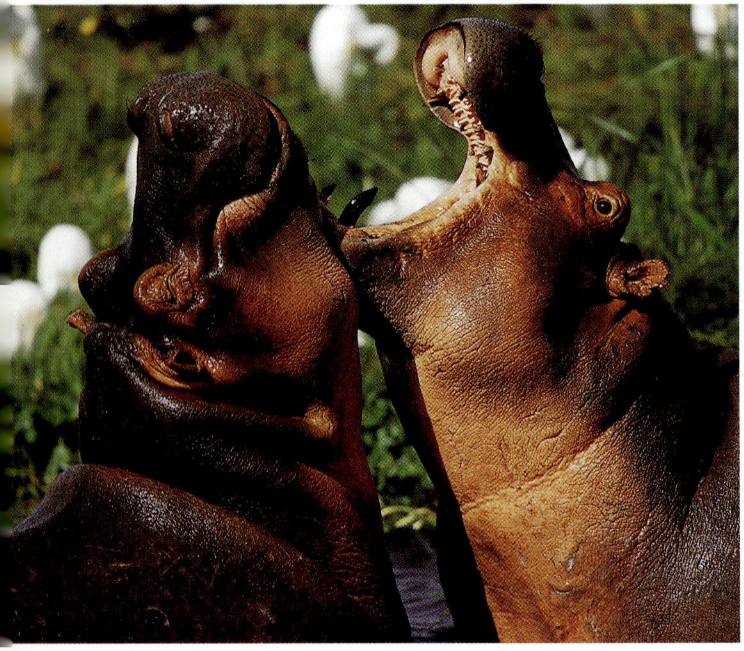

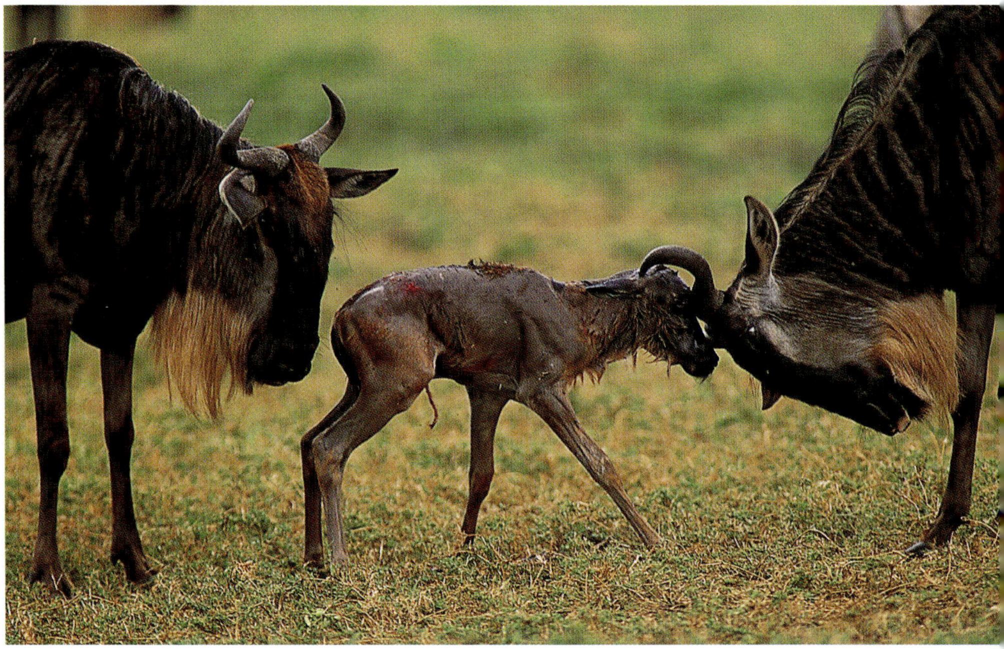

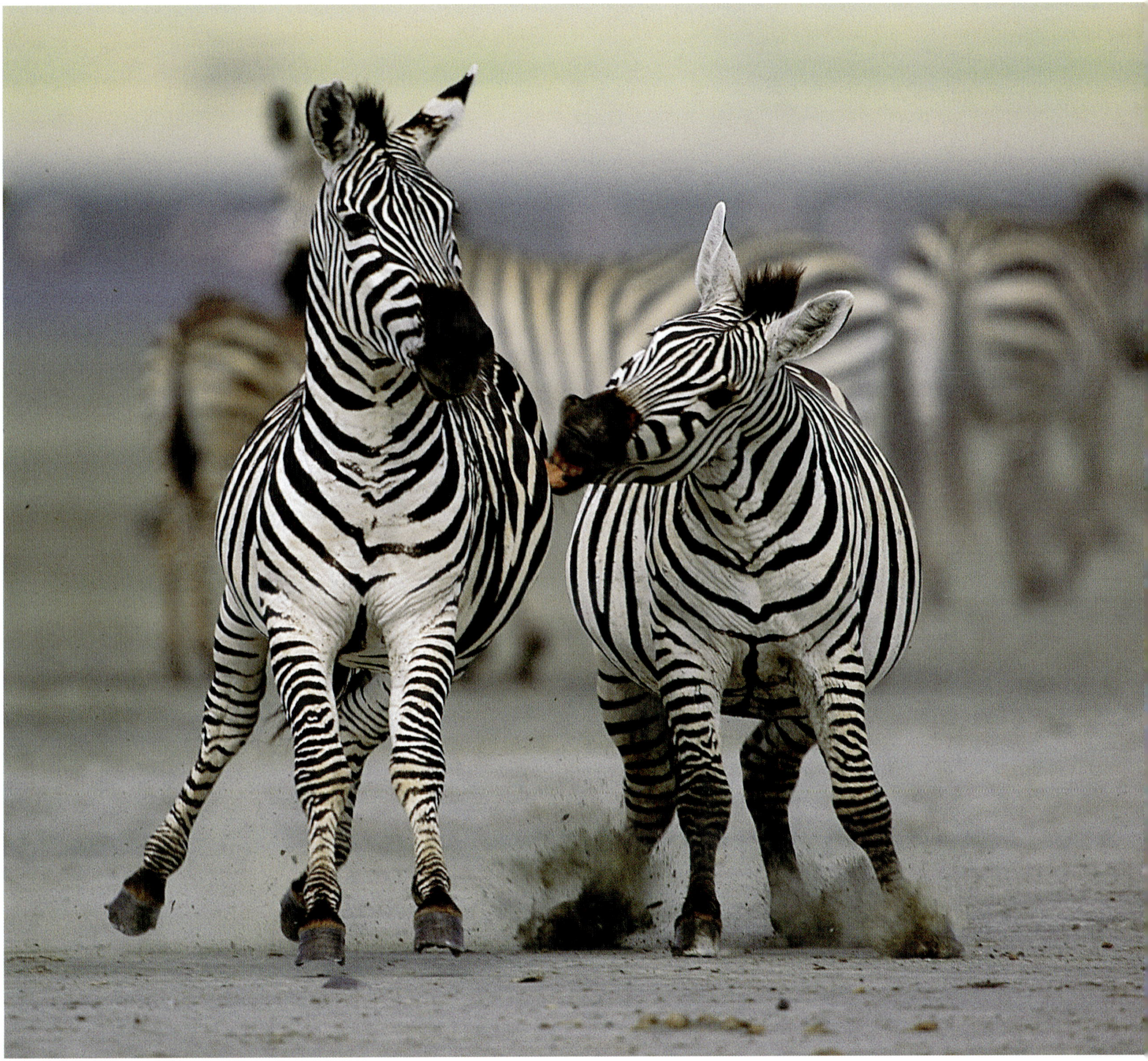

Right: A zebra foal in the crater waits out the rain within sight of its grazing mother.
Below: Traversing the plains inside Ngorongoro Crater for edibles, a Kori Bustard, Ardeotis kori, is caught in an afternoon downpour. The crater is a good place to see large birds, including storks, cranes and ostriches.
Opposite: After heavy rains, parts of the Crater Highlands are carpeted in purple flowers.

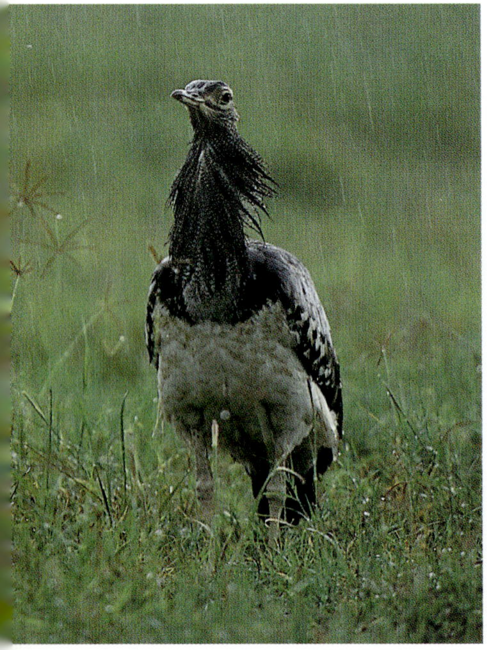

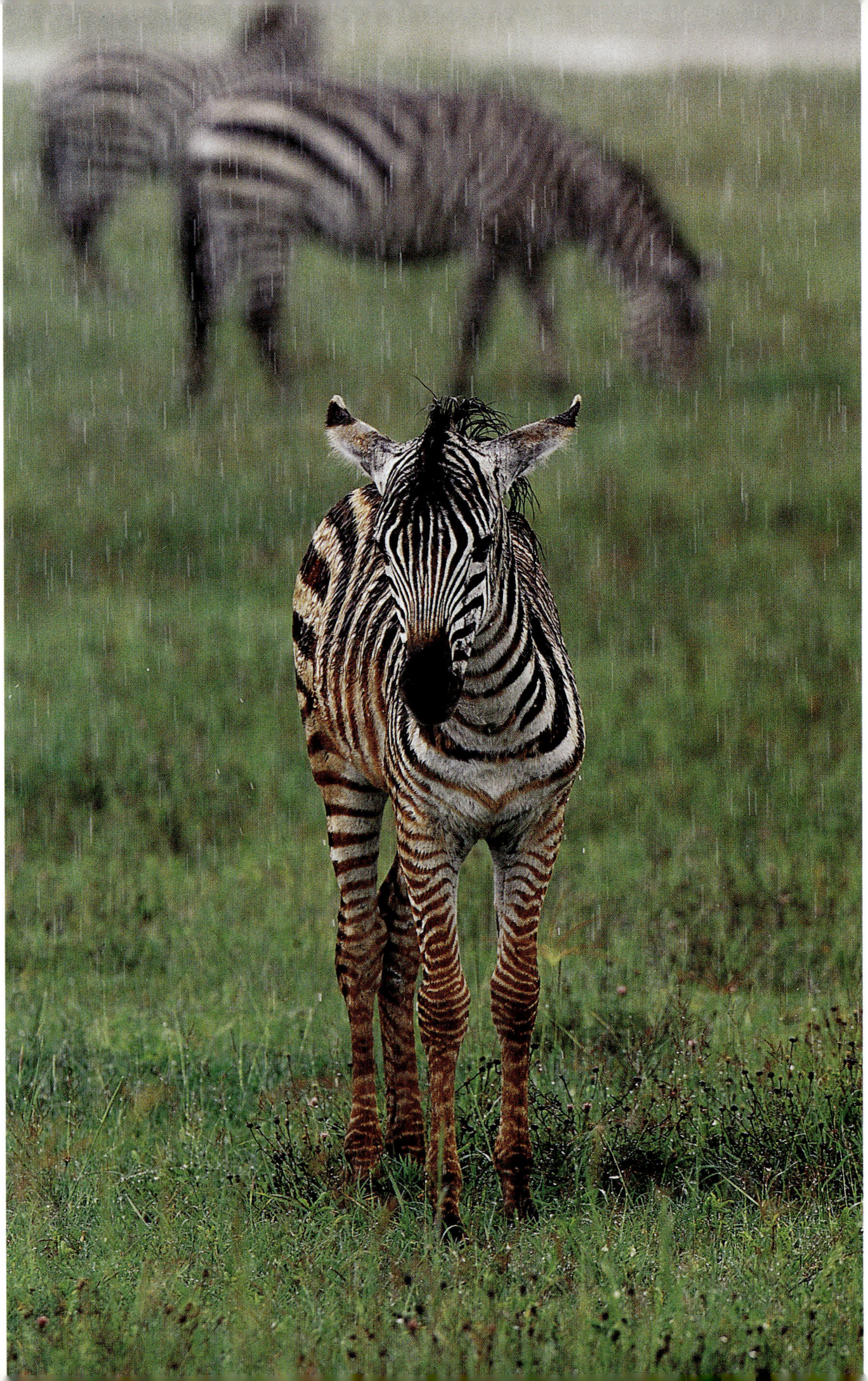

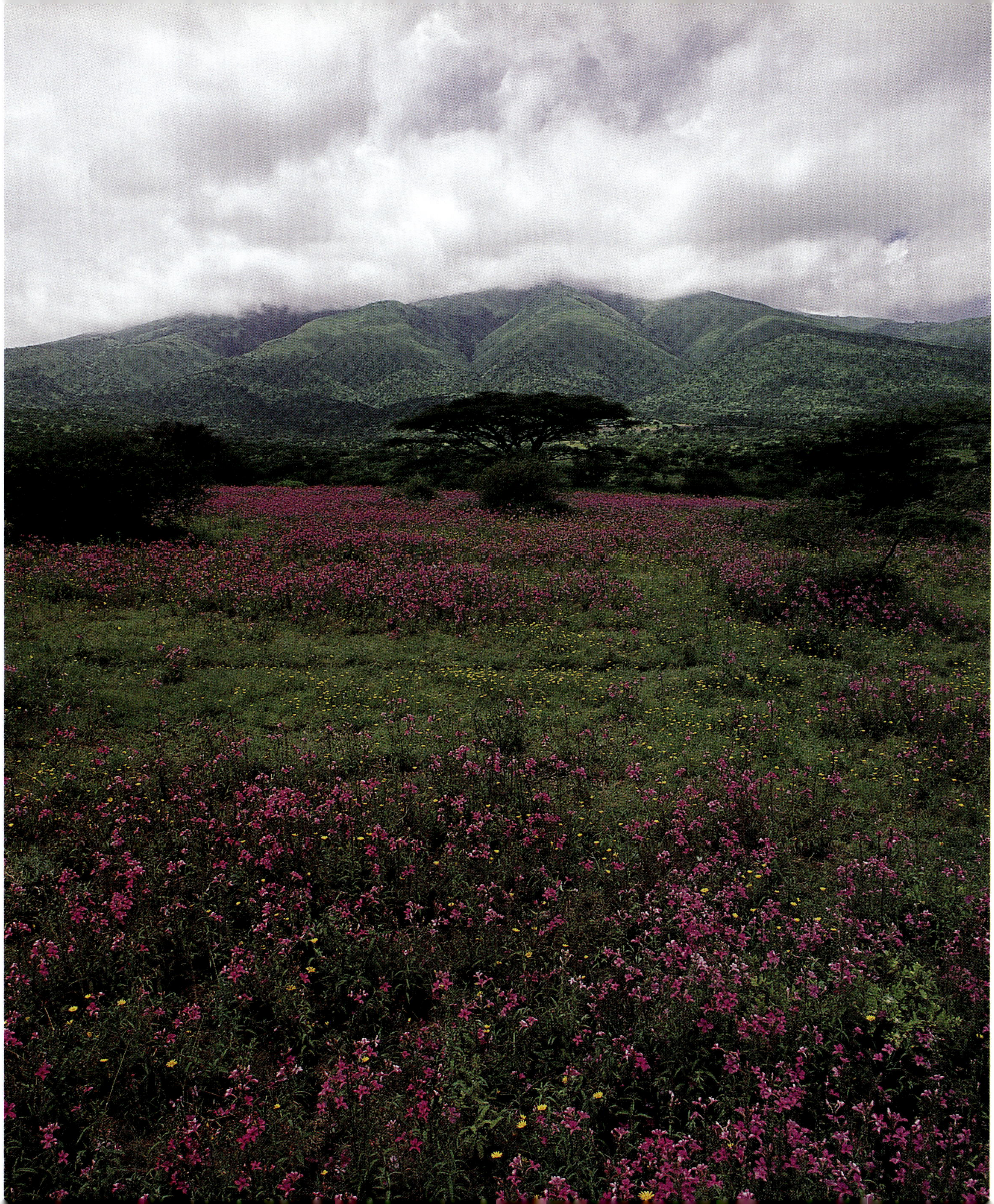

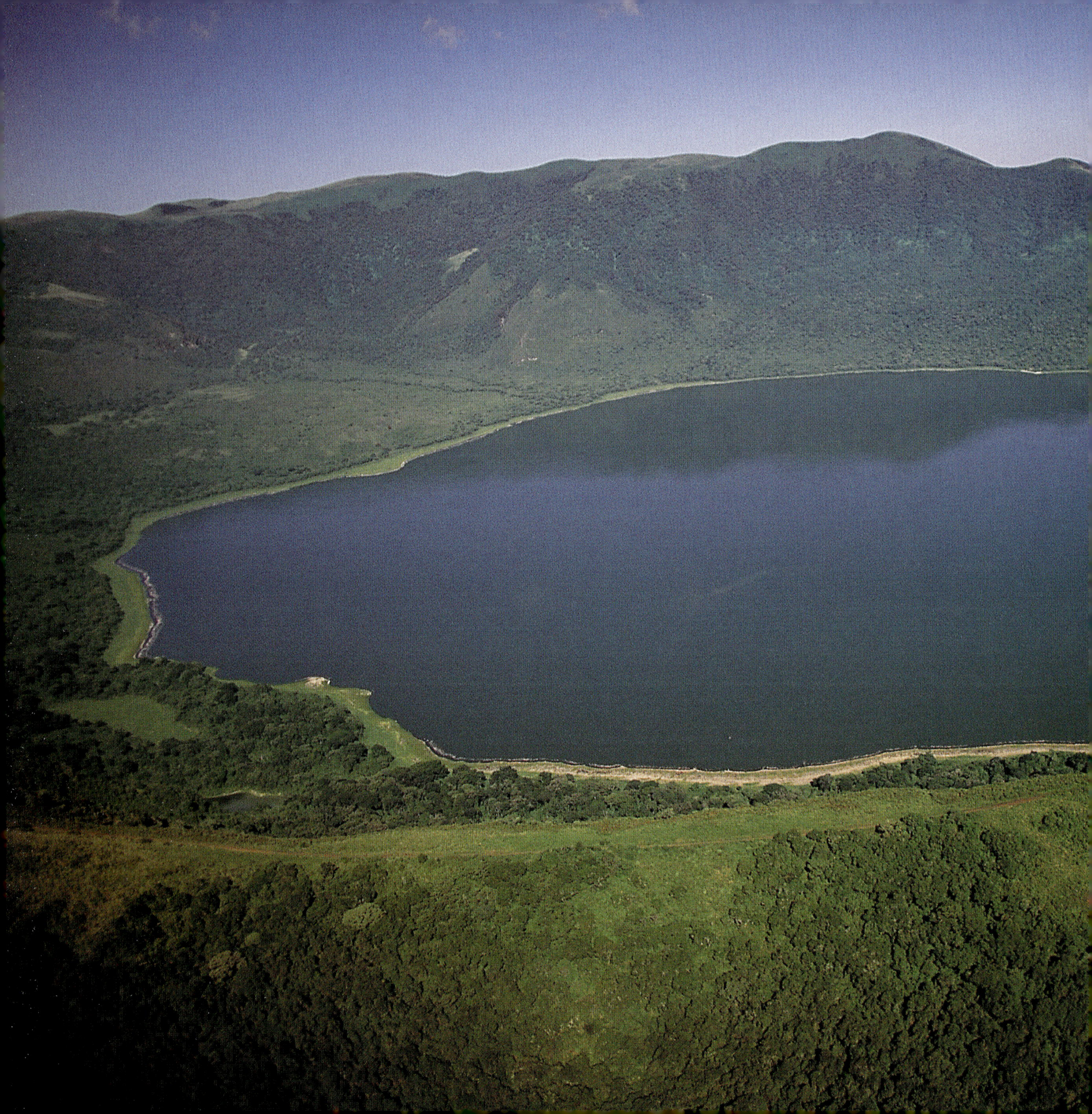

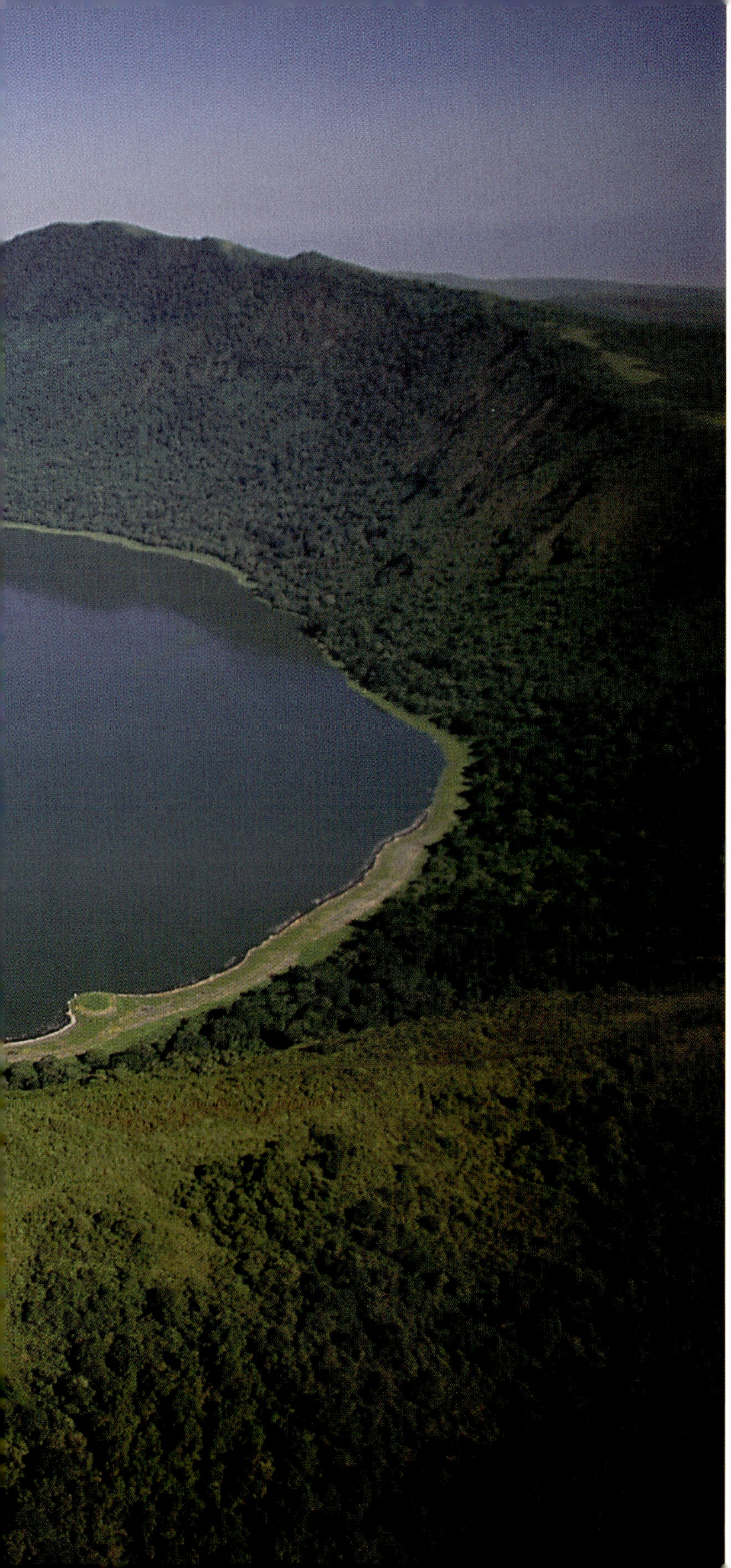

Beyond Ngorongoro

There is wildlife in other parts of the Crater Highlands, but nowhere is this as abundant as in Ngorongoro Crater. One of the reasons for this abundance is that a large section of the crater floor is covered by plains and serviced by perennial water. In the Ngorongoro Conservation Area, within which the Crater Highlands lie, there are other extensive areas of plains – around the Gol Mountains, Ndutu, Laetoli and the Olduvai Gorge – but they all lack a permanent source of water because they lie in the rain shadow created by the highlands. For this reason, the wildlife of the plains beyond the highlands is seasonal.

Past lives

There was a time when wildlife was abundant in the area encompassing Laetoli, Ndutu and Olduvai Gorge. A lake is thought to have existed in the area, and to have expanded and contracted over time, before finally drying up. At the time when the small, upright apes laid down their footprints at Laetoli, some 3.6 million years ago, they shared the gorge with other animals; the footprints of some of these animals have also been preserved in the volcanic ash from Sadiman Volcano. Over a relatively short period, Sadiman erupted some 18 times; during each event, the raining ash preserved another selection of animal prints. The largest ash layer to be exposed – named 'Pompeii' by geologists – has revealed approximately 18 400 tracks, including rhinoceros, pig, antelope, large cat, ostrich, hare, guineafowl and insect prints. Other exposed layers have impressions of elephant, buffalo and giraffe tracks.

Even then, the ecological model of feline and canine predators and their grazing prey appears to have been well established. The niches that exist today, existed then and were occupied. At that time Laetoli would have presented a similar landscape to that of today, but with a slightly different cast: the cats would have been slightly larger and sabre-toothed, and the main pack hunters would have been hyaenas as large as the lions that prowl the plains today. The famous Serengeti plains to the west had yet to be created by Kerimasi Volcano, then still being formed by subterranean geological forces.

The lake inside Empakaai, one of the tallest volcanoes in the Crater Highlands, attracts flamingos from nearby Lake Natron.

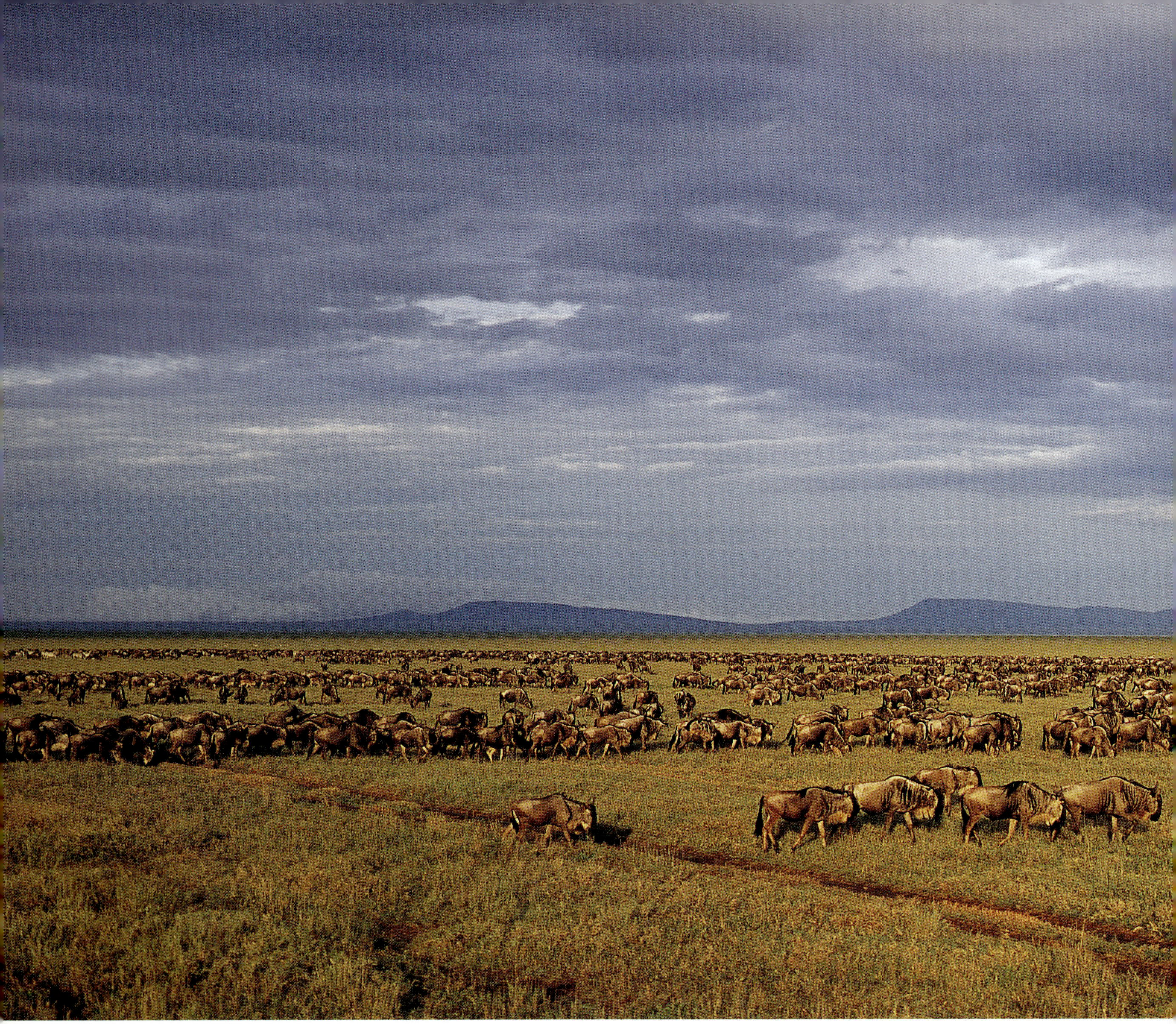

Wildebeest, Connochaetes taurinus*, on the move across the Serengeti short grass plains.*

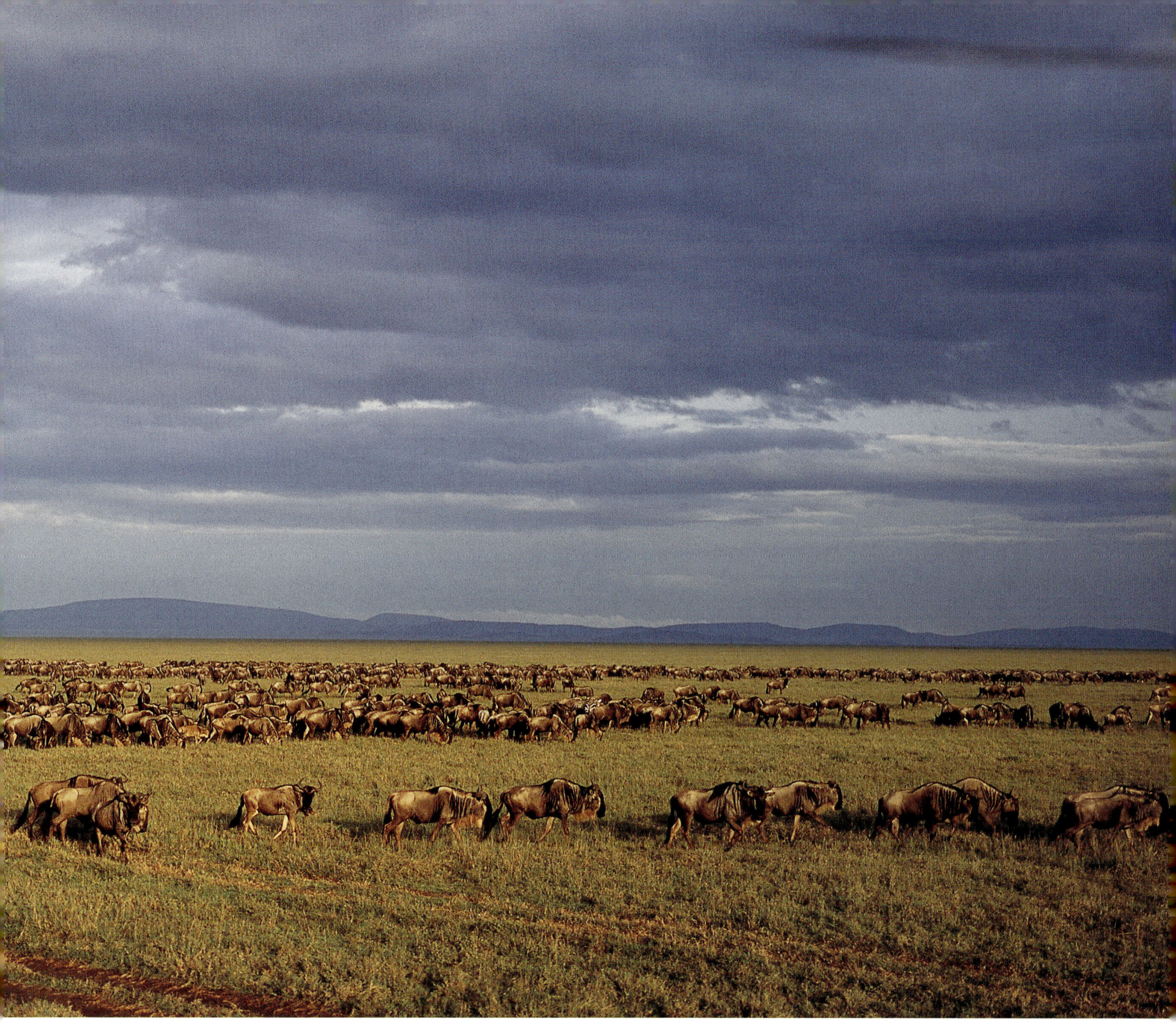

4 Migrants of the ash-covered plains

The Serengeti plains

As you leave behind the Ngorongoro Crater, taking the road west to Serengeti, the landscape suddenly drops away to reveal an immense open space. As the altitude of the descent continues to drop steadily, the surroundings become more and more like plains. The lushness of the vegetation gradually diminishes. The tall forest tree, *Acacia lahai*, gives way to the lower form, *Acacia tortilis*, of this typically African tree family. The land becomes dustier and more arid, and the moist chill of mornings on the Crater Highlands gives way to the dry heat of the lower altitudes. Stretching to the horizon and beyond, treeless, seemingly endless, lie the short grass plains of the Serengeti.

The ash-covered plains have a shallow layer of soil, beneath which is a core that is as hard as cement. Consequently, trees and bushes are largely absent. The short grass plains are dotted with kopjes – hard granite rocky outcrops exposed by the erosive action of rain and wind.

Formation

The magnificent Serengeti plains, epic in proportion and home to one of the greatest concentrations of large mammals on Earth, are formed mainly from the ash that was blown over from the Crater Highlands to the east. The principal source of the light-brown, fine-grained ash is Kerimasi Volcano in the Crater Highlands. Two million years ago, Kerimasi was responsible for an outpouring of life, the product of the alchemy of lava and water. The lava, rising up from an opening in the mountain, was thin and very alkaline. For years it poured into its mountaintop

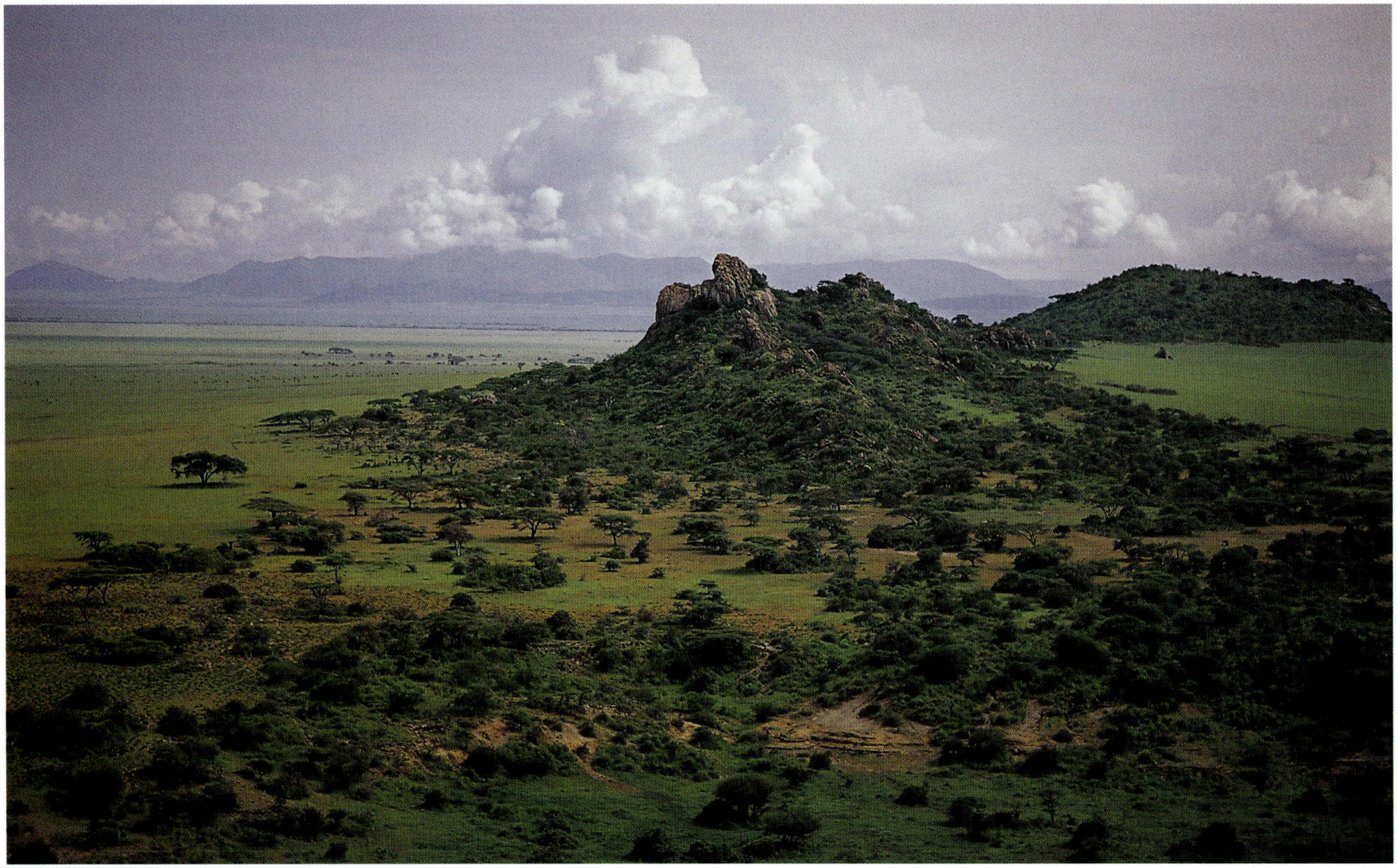

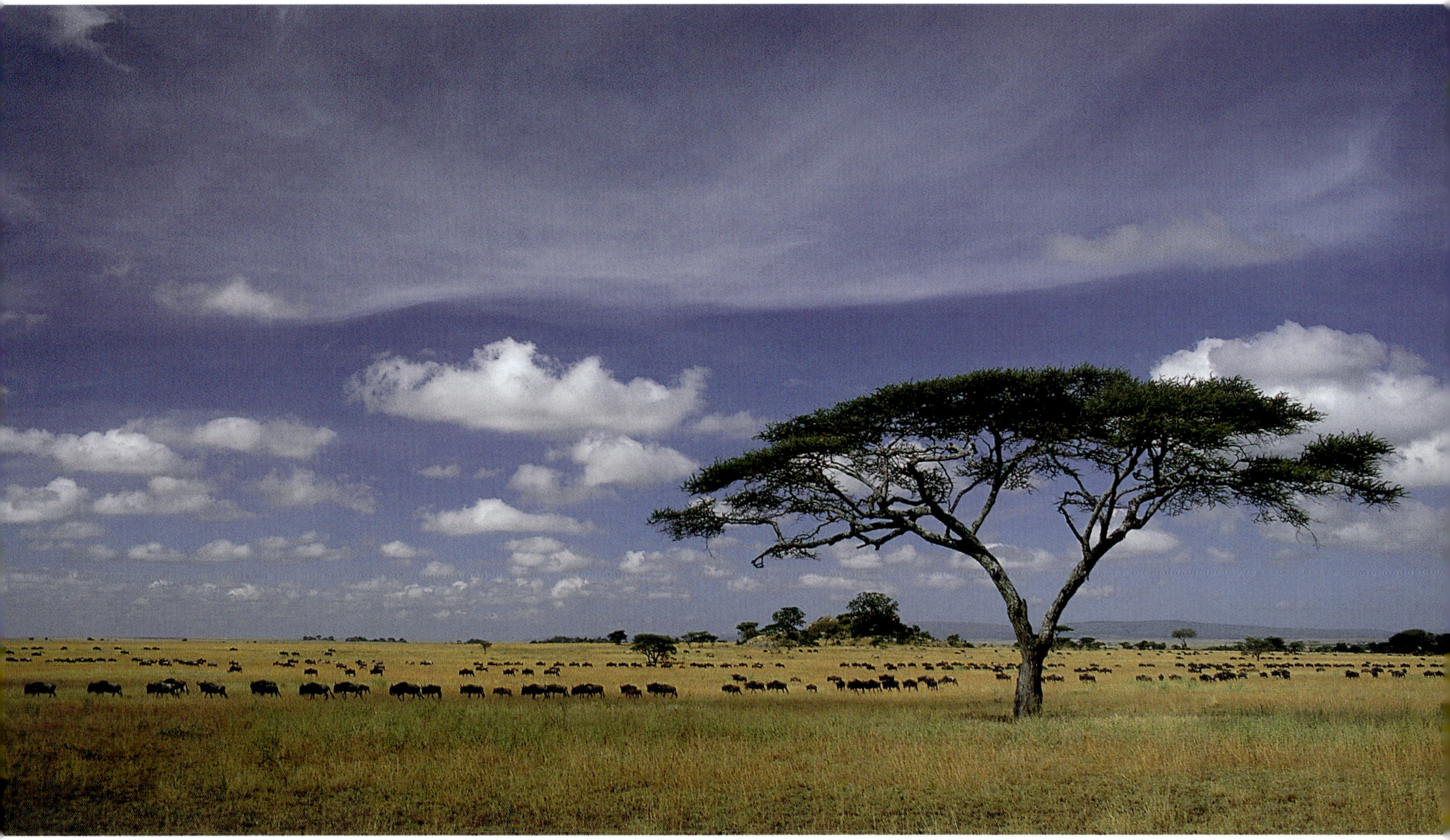

cauldron until, filled to the brim, the cauldron collapsed and the lava plunged back into the mountain, coming into contact with a vast underground reservoir of water. The clash of temperatures caused a massive explosion, instantly transforming the lava into a fountain of ash that spread over the rift valley, where the winds from the Indian Ocean pushed it to the west. The heaviest ash fell back on to the volcano in caustic drifts of sodium carbonate. The lighter and more finely textured ash, swept along in grey dunes, extended as far as 65 kilometres (40 miles) from the mountain base. Then, at the edges of the dunes, the ash mixed with water and settled into the soil.

The wildebeest migration arrives in Seronera Valley in central Serengeti at the onset of the long, dry season. The area has permanent water sources.

Under the surface, some of the ash formed impenetrable tuff, its grains compactly bound together; these concretions today impede tree roots seeking deeper purchase. Moreover, the ash at the surface is so porous that its high calcium content is leached to underlying layers a metre below the surface, where it forms a pavement of calcium carbonate harder than concrete, which also obstructs the probing roots of trees. As a result, grasses – mainly dwarf forms of drop-seed and star grass – exploit this niche unhindered by trees, and flourish abundantly and extensively.

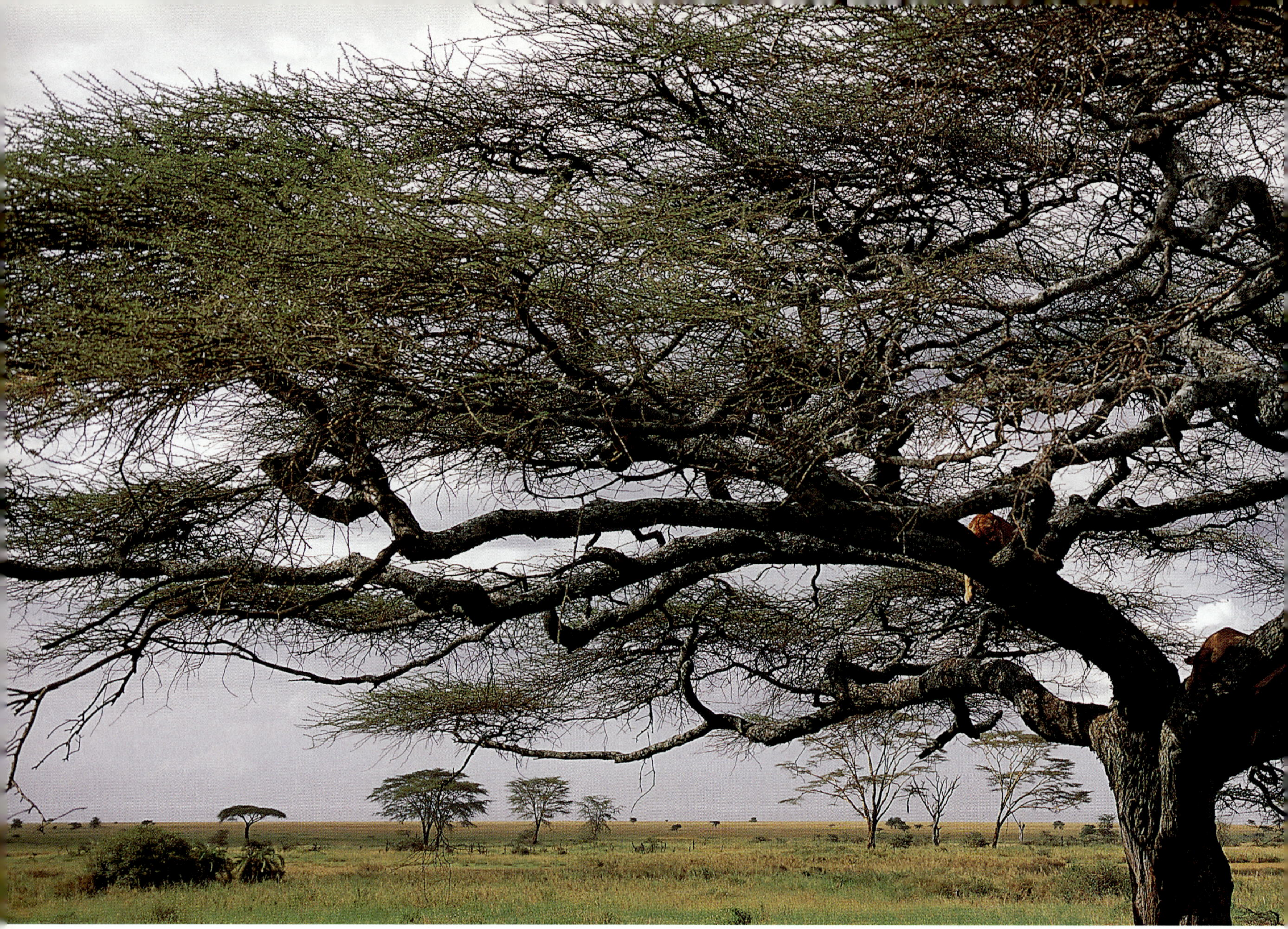

Soil, grass and water

The porous ash does not hold water for long periods. Without rain – the dry season lasts from May through to October – the grass is of no nutritional value, and cannot support grazers; there is little animal life on the short grass plains. The Crater Highlands are responsible for this scarcity of moisture: they block the moist, easterly winds off the Indian Ocean for most of the year, leaving the western slopes in a barren rain shadow – a region from which rain-bearing clouds are excluded. Further to the north and west, close to Lake Victoria, rain falls during most months of the year, but the soil there is leached of vital nutrients.

When the inter-tropical convergence passes through East Africa each year, the monsoon rains build up in intensity until, in November, they have finally amassed sufficient power to overwhelm the terrestrial barrier of the Crater

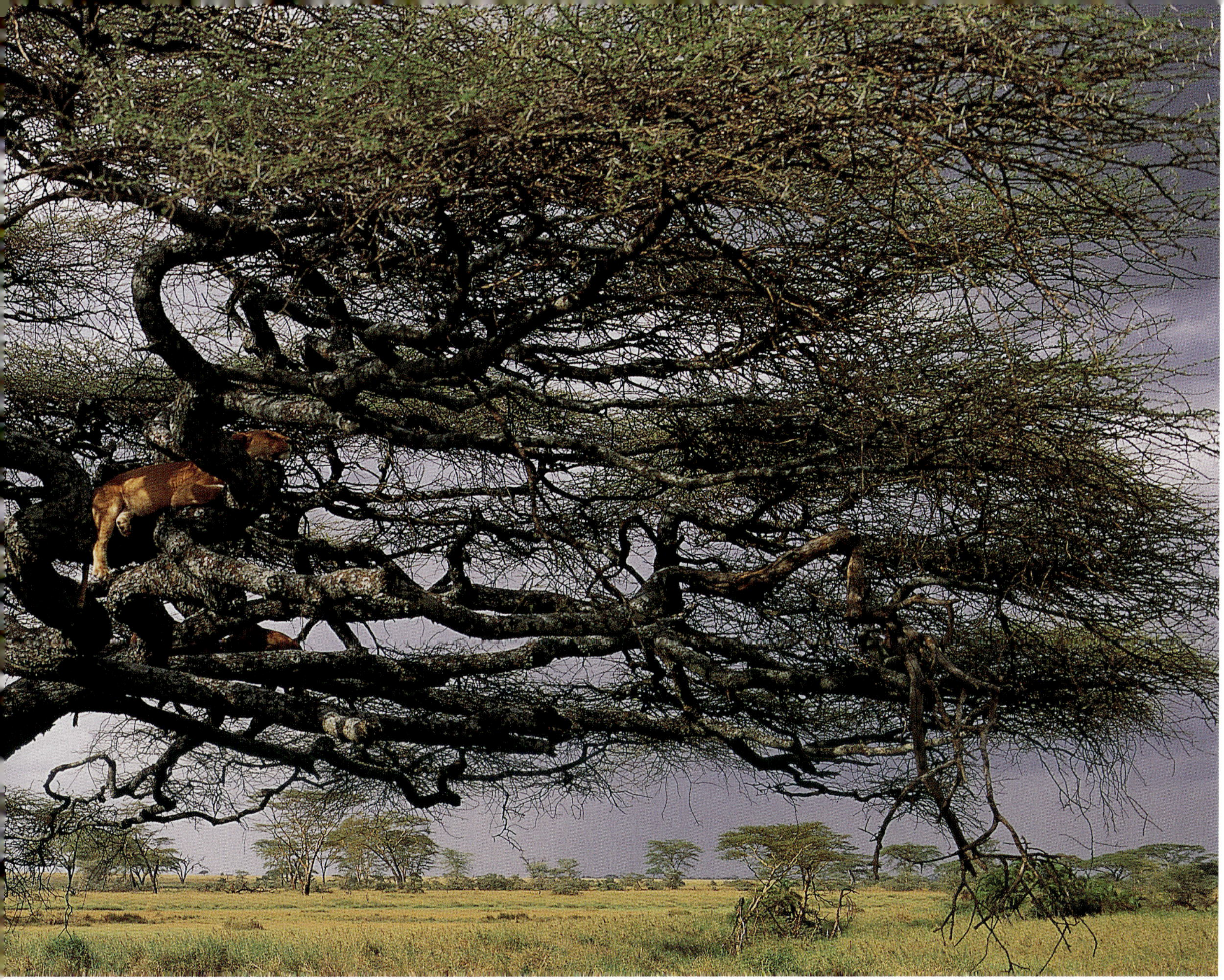

Highlands. With the clouds come the migrant grazers – two million strong – as the desert becomes a luxuriant meadow. Vast herds of zebra, wildebeest and Thomson's gazelle are drawn here between November and April by the quality of the nutritious grass. The eruptions of Kerimasi two million years ago have ensured that the soil today remains rich in minerals: sodium, calcium, potassium and phosphorous. The grass is easy to crop – it does not shoot up, lank and unpalatable, when it rains – and can be grazed again and again, because it recovers fast after each cropping.

Nowhere on the plains do the rains fall evenly or predictably. The great herds have adapted to this uncertainty by rotating around the pastures, frequently on the move, following the rains and the grass – the twin elements on which they depend. The rains encourage the grass and affect its growth; this growth regulates the numbers of herbivores the land can support; the herbivores, by their abundance, dictate the numbers of predators.

Seronera's tree-climbing lions, Panthera leo, *take a siesta high up in a tree, where the breeze is cooler; the tree also serves as a vantage point.*

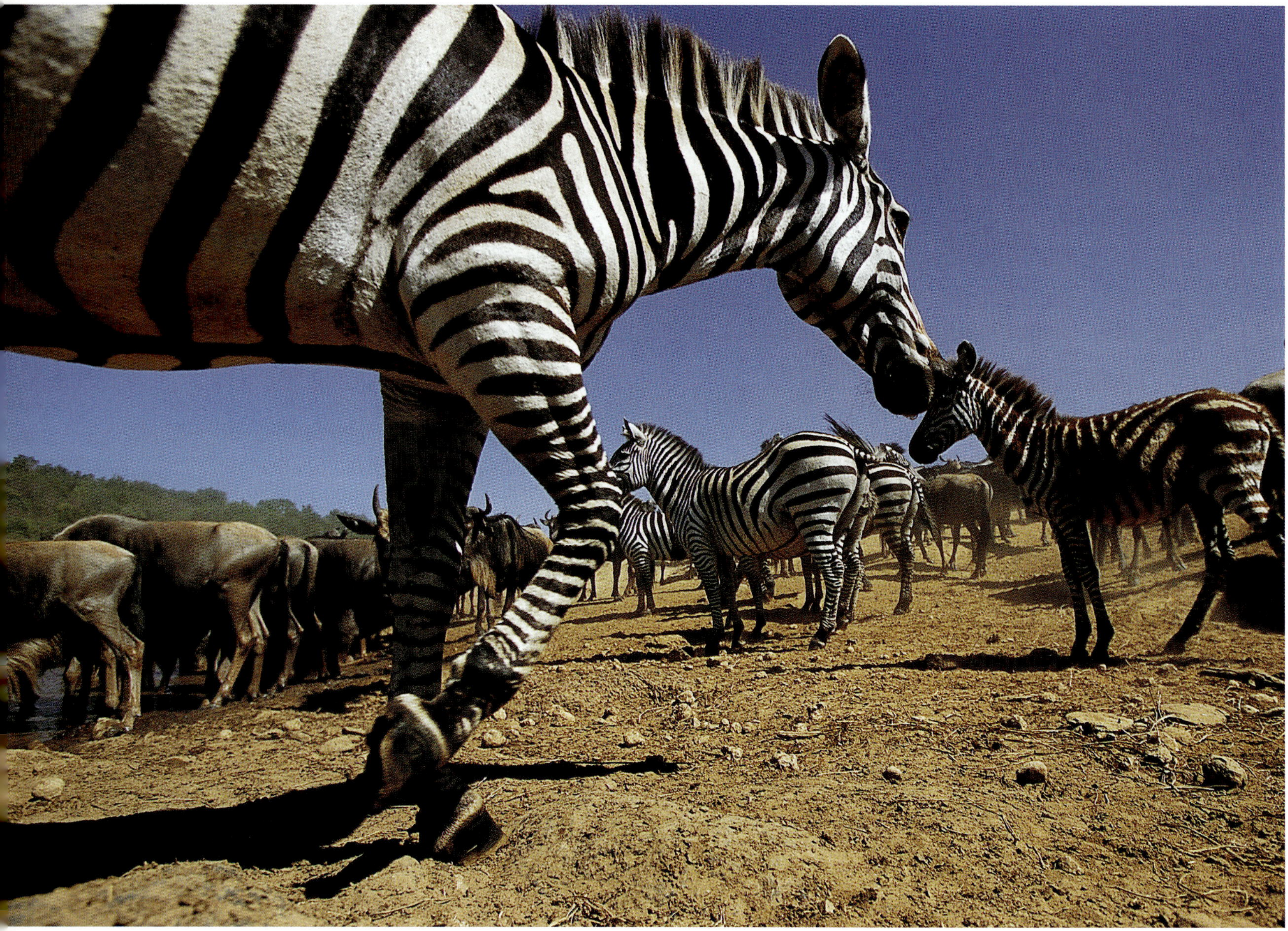

Migratory herds of wildebeest and zebra, Equus burchelli, *gather at the bank of a river as the long, dry season bites.*

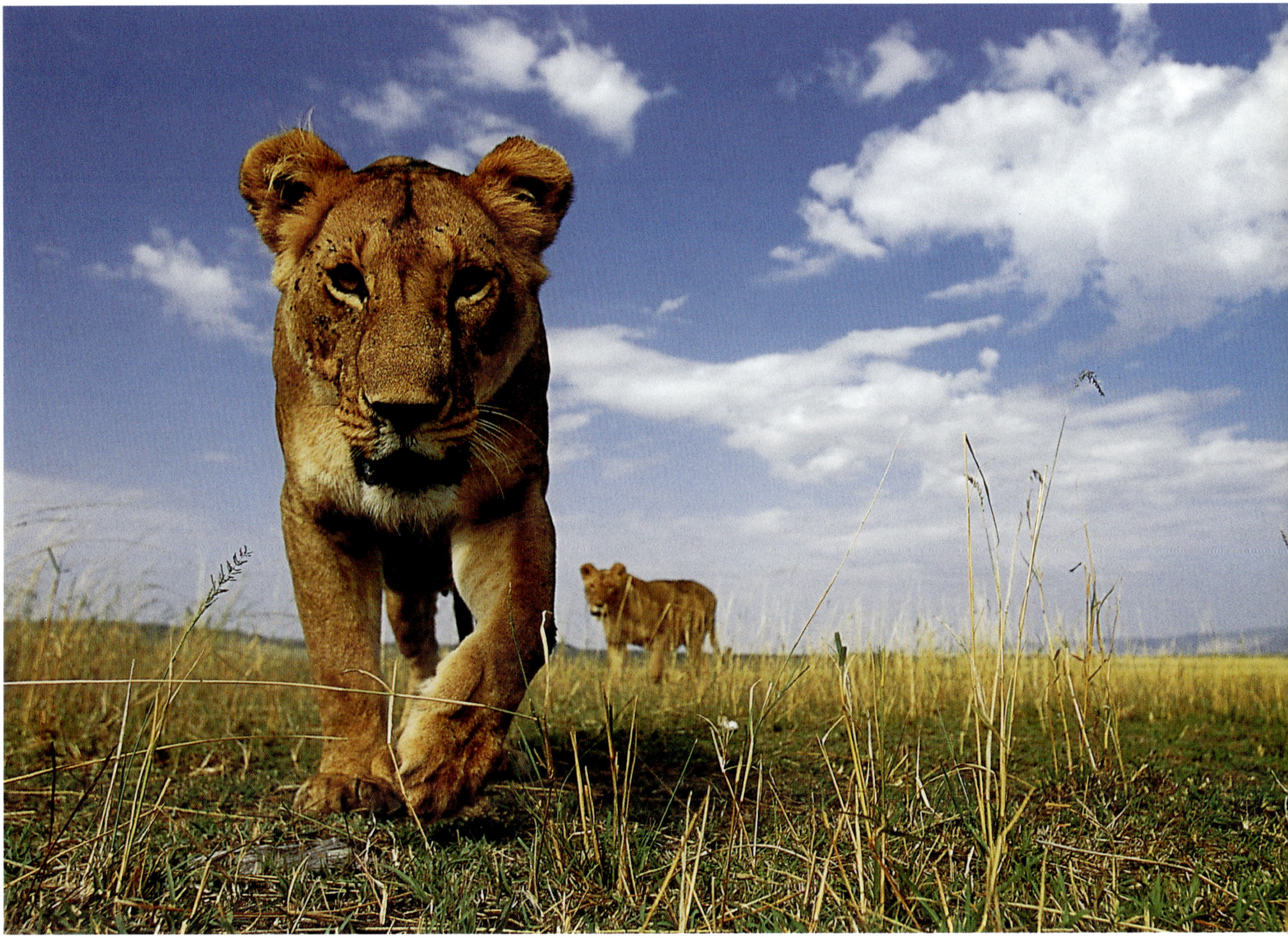

Above: *Two of the many nomadic lions that follow the migratory food supply.*

Following spread: *Sandgrouse,* Pterocles exustus, *arrive early in the morning at a waterhole in the Serengeti. Males wet their feathers to carry moisture back to their chicks.*

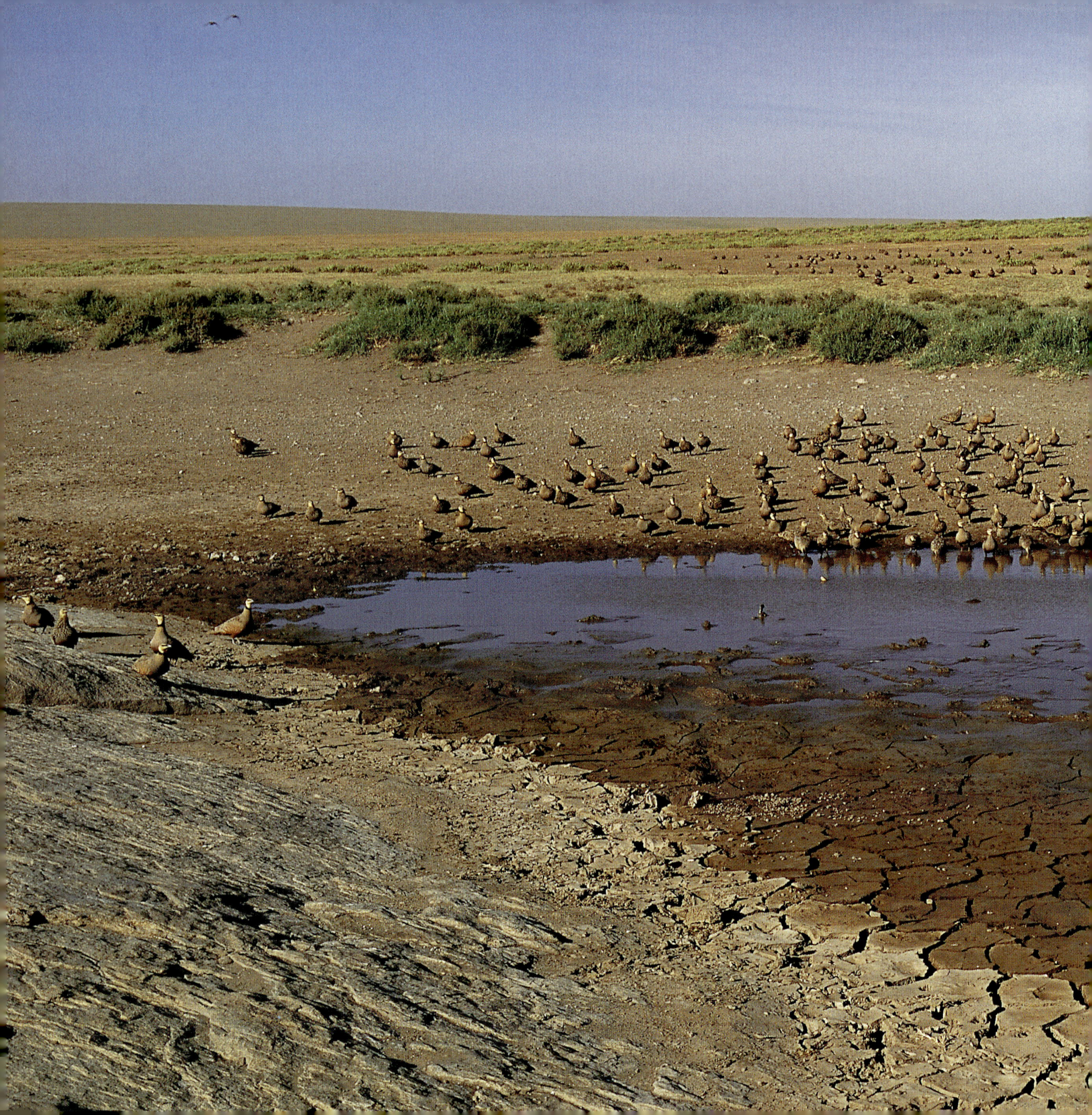

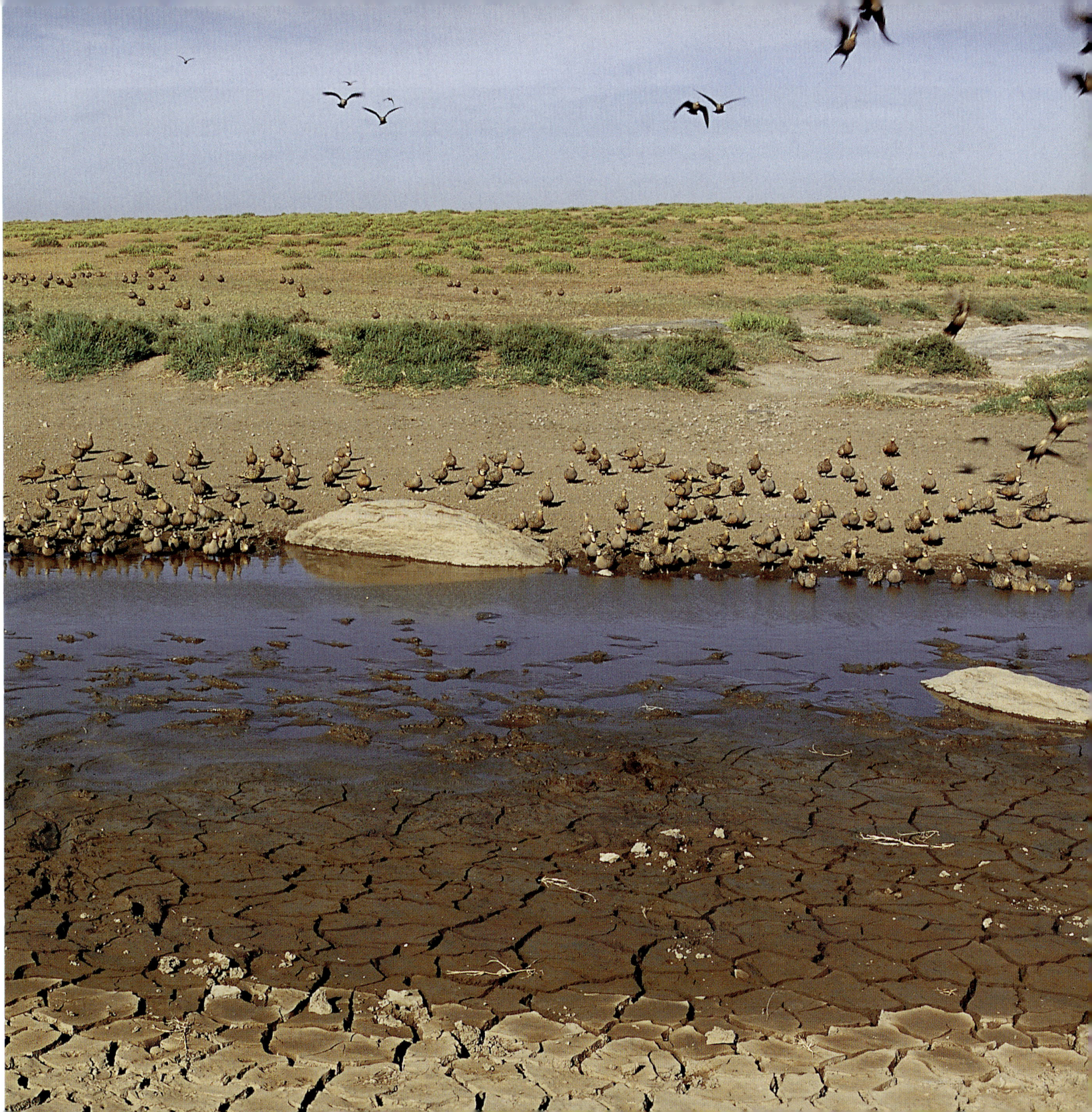

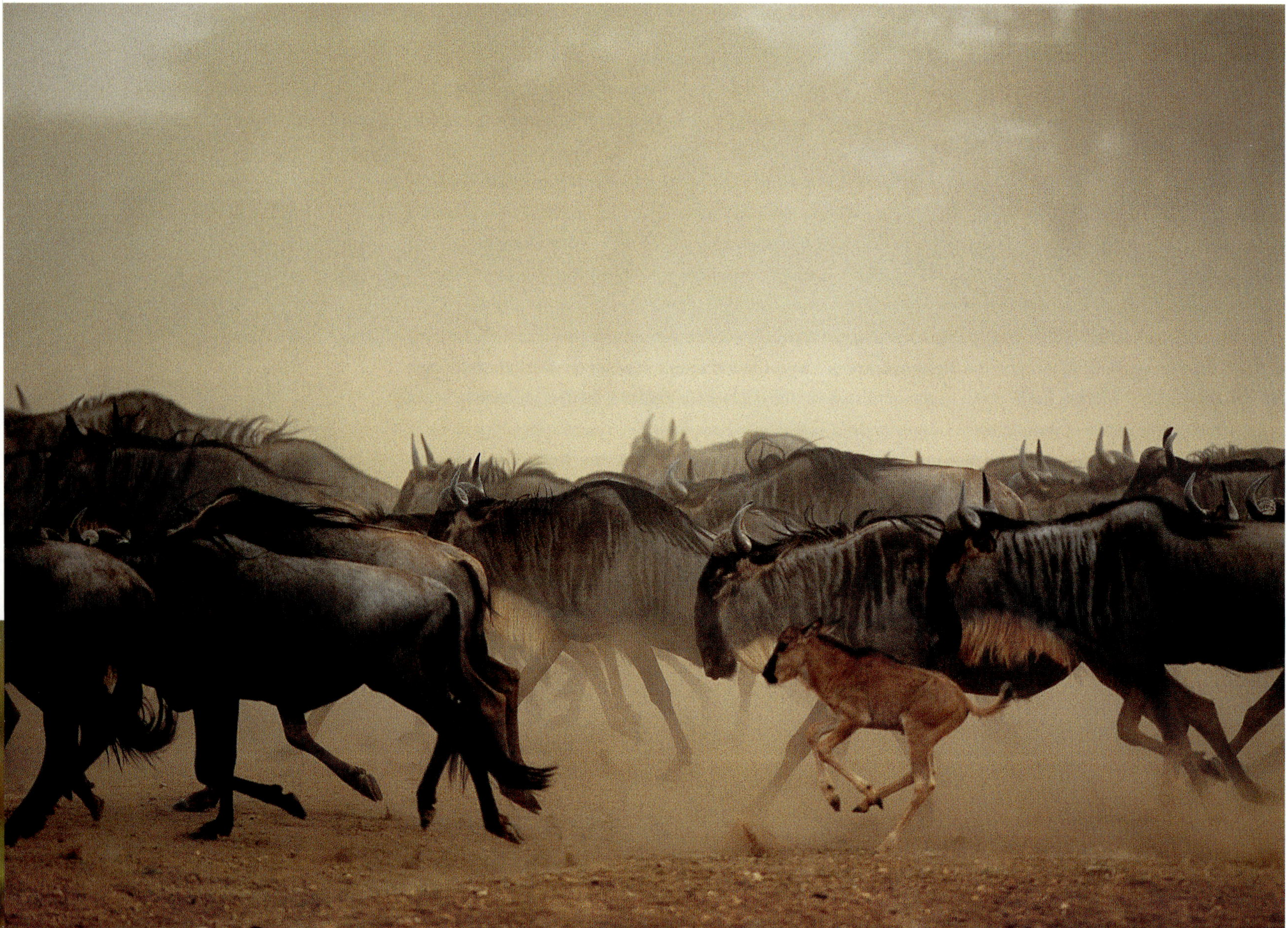

A wildebeest calf keeps up with the fast-running herd. Within two weeks of birth, a calf can run almost as fast as its mother.

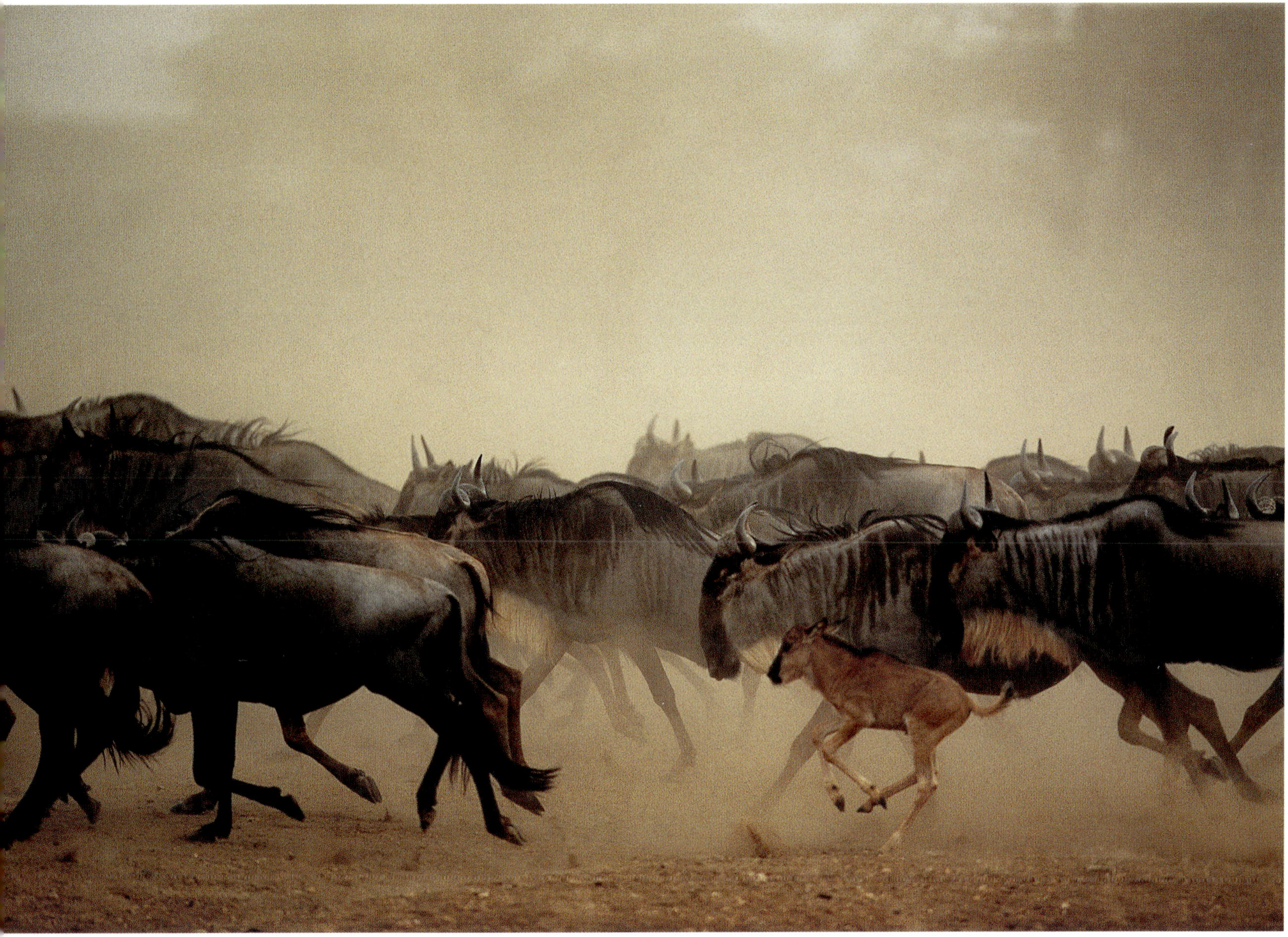

A wildebeest calf keeps up with the fast-running herd. Within two weeks of birth, a calf can run almost as fast as its mother.

The remaining waterholes become increasingly foul, and eventually dry up completely; the land shrivels without moisture. The zebra and the wildebeest start to leave the short grass plains, bequeathing them temporarily to Thomson's gazelle, which linger awhile before bringing up the rear in the great migration to the long grass plains, and then onward to the woodlands. By migrating northward and westward, where there is grazing and water, the herds bridge the seasonal variation in their food supply.

The herds move through the long grass plains of central Serengeti during the months of May and the first half of June, and some then venture into the Western Corridor. In terms of vegetation, much of the Western Corridor is dry, thorny country. Here, under stony hills, there are also acacia woodlands and open plains dotted with spiky thickets of wild sisal. The combination of grasses found in the Western Corridor reflects the impact of seasonal fires, which for centuries have swept through these areas during the dry season, thinning out the woodlands and enabling particular fire-adapted grass species to evolve and thrive. Permanent woodlands demarcate the position of the Grumeti River, flowing through the Western Corridor – a year-round source of water on which the migrating animals depend, but which is not without its dangers: some of the migrants will fall prey to river crocodiles, which can reach six metres (20 feet) in length.

The herds that migrate northward also have to cross the Grumeti River. Just as in the Western Corridor, the soils of the northern Serengeti are deep and ancient, supporting longer grass roots and taller grass stems. The herds must traverse woodlands for part of the trek in order to reach the nutritious grasslands of the northernmost part of the Serengeti. The woodlands present challenges to the herds: they harbour parasites, which follow on the wing, or travel inside the flesh or fastened to the skin of passing animals. With the parasites come diseases that take advantage of the stress caused by the dry conditions. But since migrating northward is one of the few options for surviving the dry season, the herds have little choice but to run the gauntlet of these perils.

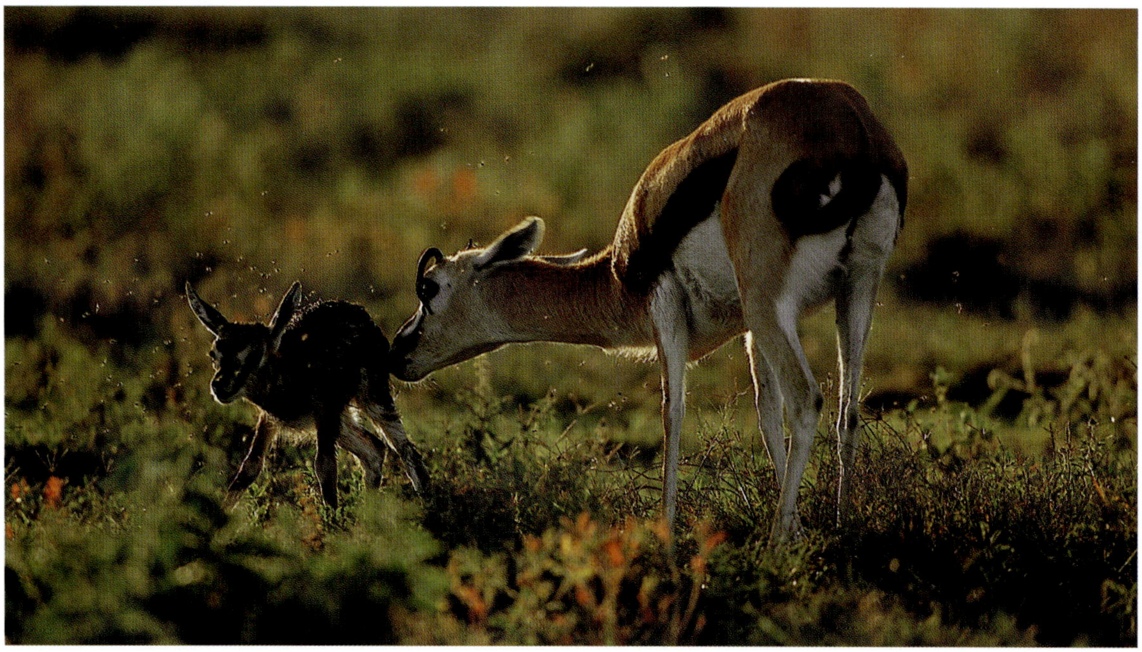

The mother of a newborn Thomson's gazelle, Gazella thomsoni*, licks the birth fluids off her newborn. After being washed, the fawn is odourless and predators cannot detect it by smell.*

The contribution of the dung beetle

The huge numbers of grazers place a heavy burden on the dung beetle, which rolls its dung into neat round balls in which it also lays its eggs. The dung beetle then buries the dung balls, so integrating the dung with the soil and preventing it from simply drying out and blowing away in the wind. It is estimated that dung beetles process more than 3.5 million kilograms (7.7 million pounds) of manure daily, cleaning up the plains and at the same time fertilising them. Dung beetles could be called the saviours of the short grass plains; without them, large tracts of this semi-arid region, with its light, friable soil, would turn into a dustbowl.

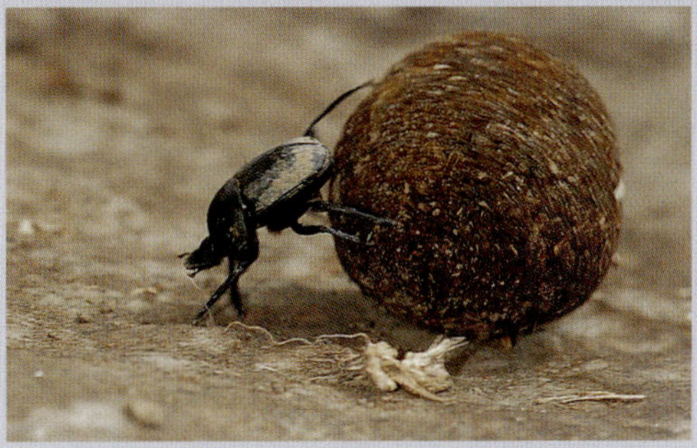

Mass calving and mass predation

Almost all wildebeest cows over three years of age give birth to single calves. During a period of just over three weeks, more than 80 per cent of all wildebeest calves for the year will be born. It is as if the herds are reproductively synchronised – almost all the calves are born at about the same time. This survival strategy is based on the principle of offering predators more than they can possibly take. Once on its feet, a calf can run almost as fast as an adult; yet despite both their abundance and their speed, a sixth of all calves are killed, mostly by lion and hyaena. In such times of plenty, with a huge surplus of available food, the predator must capitalise on the bountiful prey by filling to capacity not only its own stomach, but also the stomachs of its cubs, whose birth has been timed to coincide with the birth of the grazing animals.

A wildebeest cow drops her calf. During the mass calving in February, about 250 000 cows give birth within a period of three to four weeks.

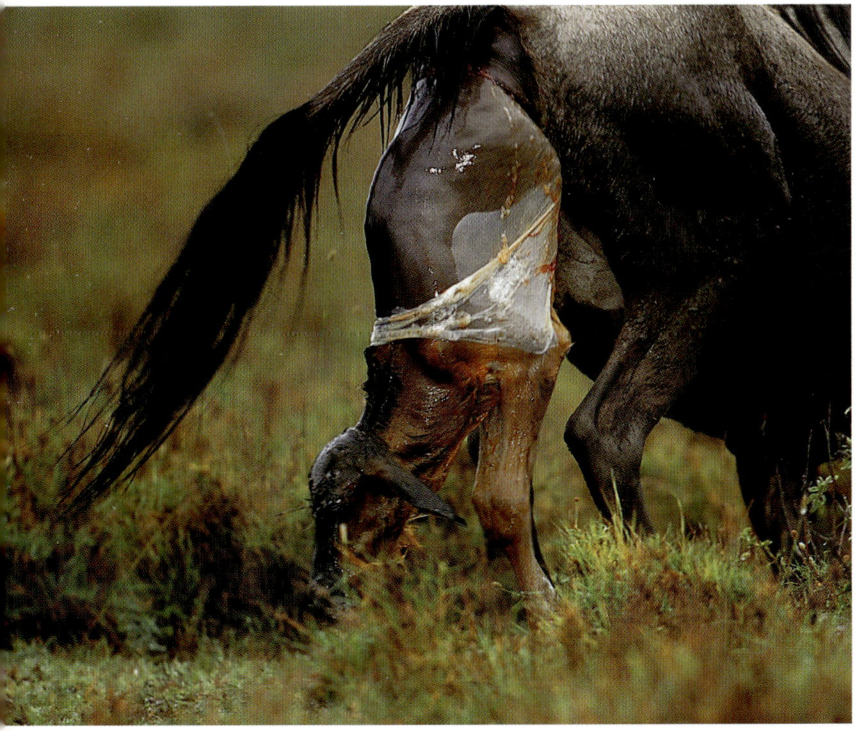

There would be more predators were it not for the limitations of the dry season. The volcanic ash is too porous to hold water for long; so, when the rain stops, a massive migration of wildlife ensues. The lion and hyaena retreat to the woodlands to eke out a living there, while first the zebra and then the wildebeest trek north and west; the further they travel, the finer the ash becomes, until it blends with other matter and starts to resemble ordinary soil. After the rains, the Serengeti short grass plains teem with wildlife; but, with the onset of the dry season, when most of the animals nurtured on the grasses depart, the plains are left severely depleted.

The great migration

In late April or early May, the wind turns hot and dry. Surface water diminishes. Heat and dust start to encroach on the landscape. A few animals well adapted to dry conditions stay on – ostrich, jackal, Grant's gazelle, mongoose, lizard and birds of prey. They survive on the plains throughout the year by making the most of the moisture in their food. As the short grass plains deliver less and less nutrition, the lactating female wildebeest are put under growing stress.

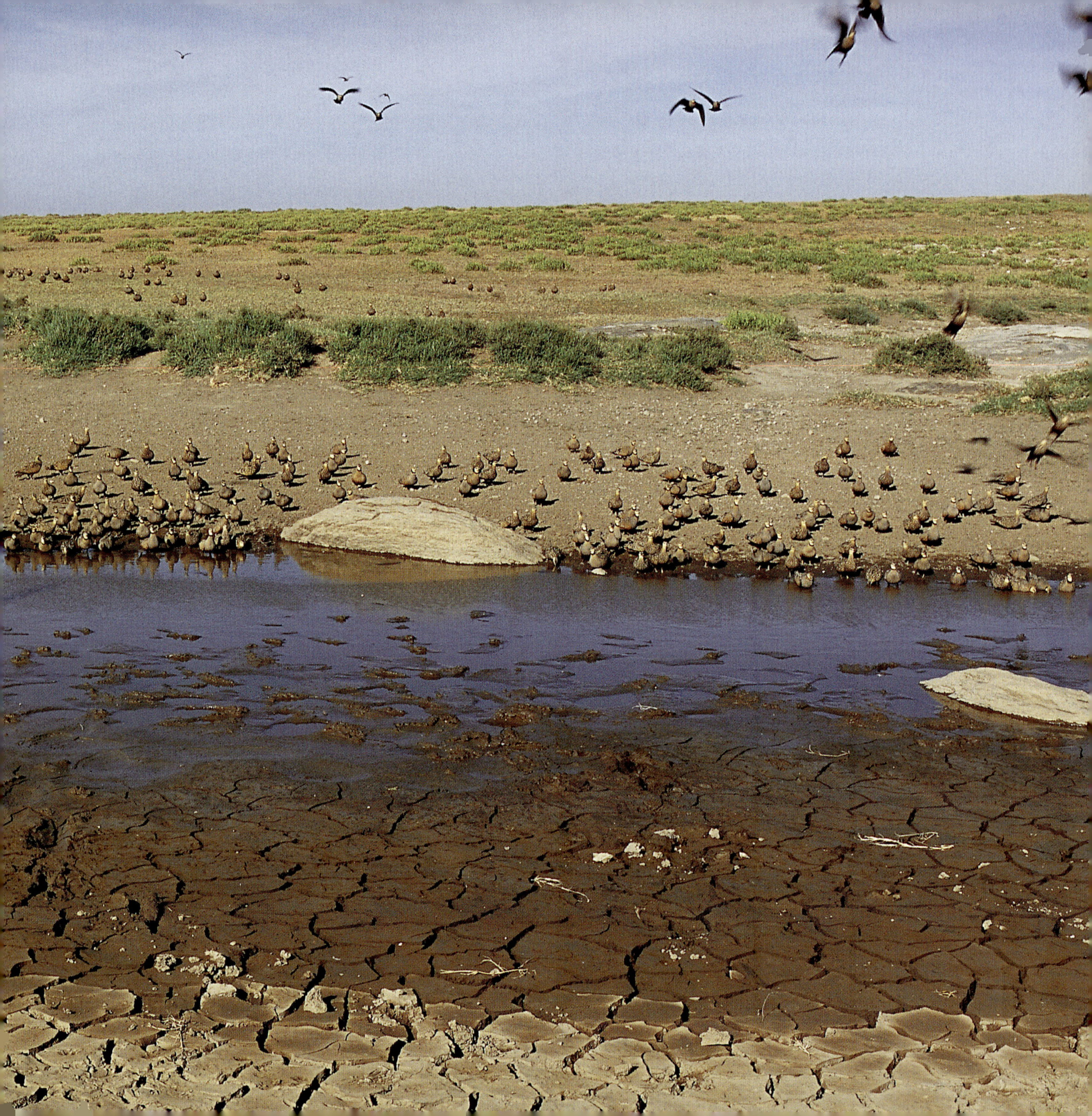

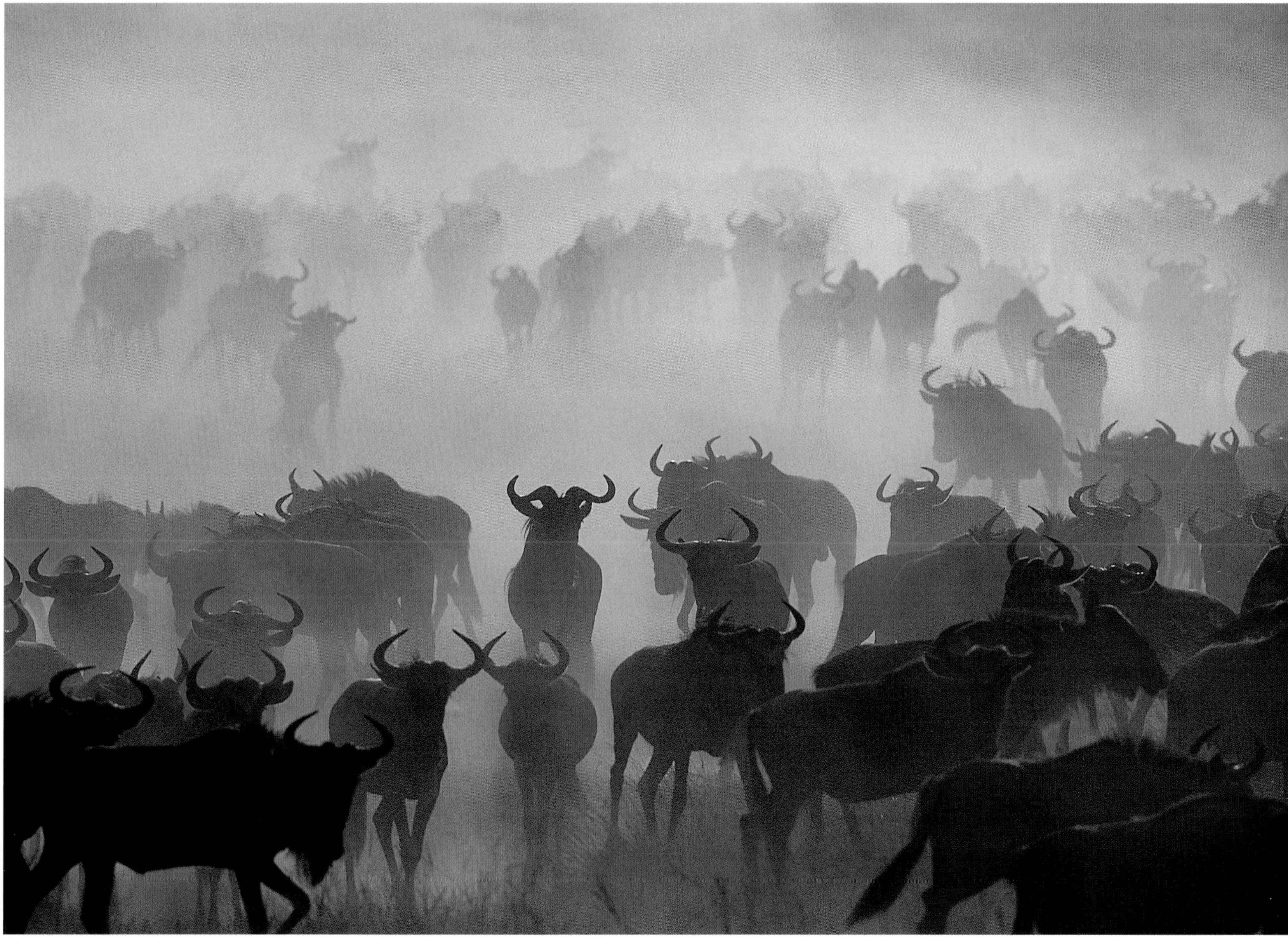

Wildebeest herds kick up dust on their march from the drying plains in the east to the more temperate lands in the northern and western Serengeti.

Crossing the Mara

The first wave of migrants crosses into Maasai Mara National Reserve in July. By the law of the seasons, they are bound to the region until the next rains. Although not far away, the rift (its eastern arm, that is) has relatively little direct influence on the Mara ecosystem, which differs from the Serengeti in having a more even distribution of rain, but which also harbours resident predators waiting for their time of plenty, when the migrants are forced to make the hazardous crossing of the Mara River.

The source of this river is a shallow swamp among the forests of the Mau Escarpment that looms over the rift valley floor. Fed by rains that fall there, the river flows throughout the year. It enters and meanders through the reserve in a

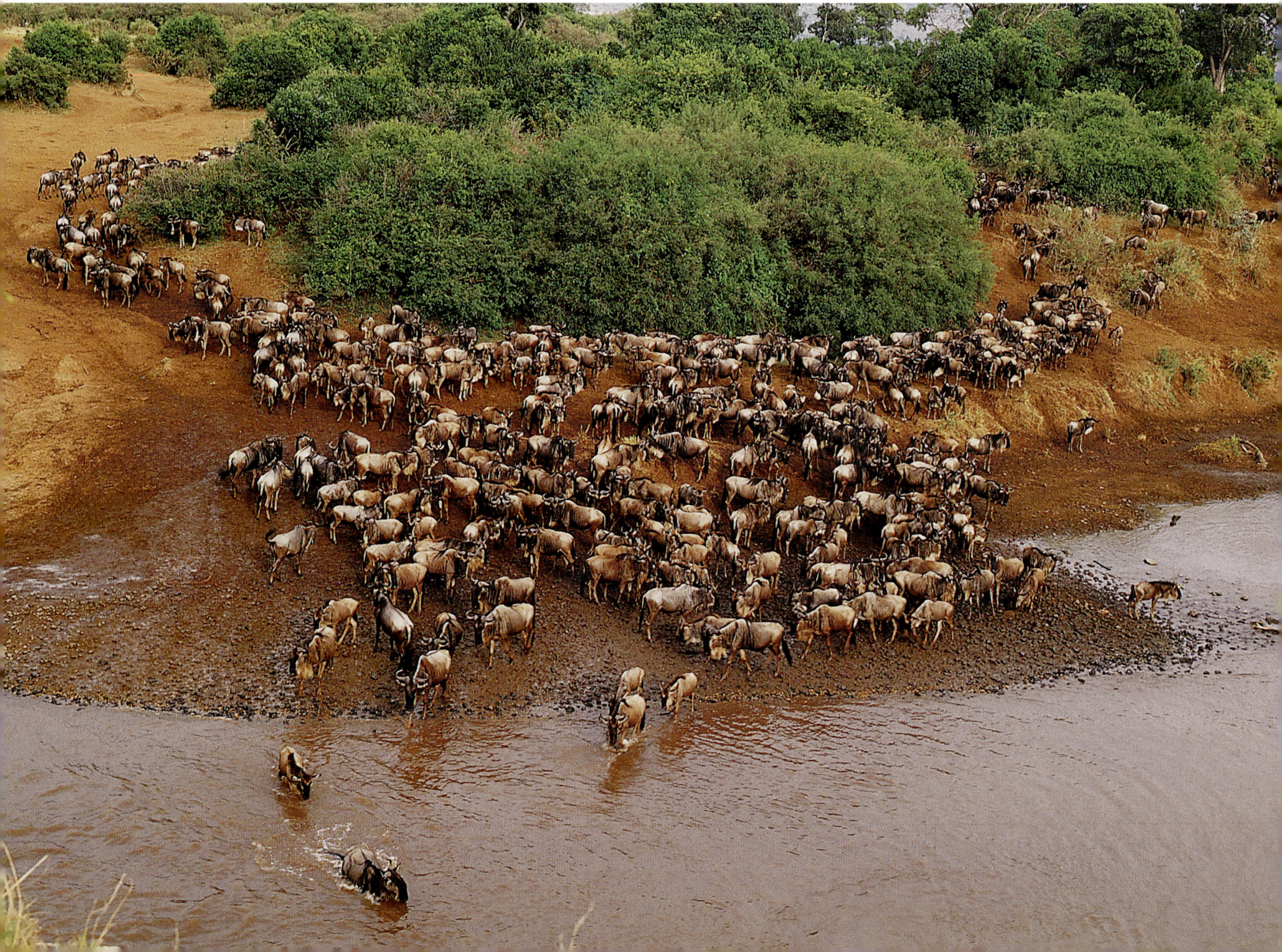

Wildebeest gather at a low bank on the Mara River in an attempt to cross to the other side for grazing. This river crossing failed on account of the steep, high bank on the opposite side, and the herds turned back to cross elsewhere.

series of loops, veering west through the northern Serengeti National Park before emptying into Lake Victoria. Since the migration moves from south to north, and back again on its return journey, the herds – whatever their route – must cross the river twice. In penetrating northward, deeper into Mara, the herds cross the river in late July and early August. The incentive to cross is the ungrazed grass on the other side – red oat grass, not as nutritious as that on the Serengeti short grass plains, but the only option for these migrants. By October, Mara is thoroughly dry, and the remaining grass is largely unpalatable. The herds start heading south when the rains return in late October or early November, and have to cross the hazardous Mara River again; this crossing, in either direction, is possibly the greatest danger they must face in the course of their great trek: an estimated one in every six wildebeest will die during the migration.

Below: *A wildebeest brings up the rear of a successful Mara River crossing.*

Bottom: *Another, having crossed the river, gets stuck in slimy mud on the opposite bank. This is one of the many hazards of a river crossing.*

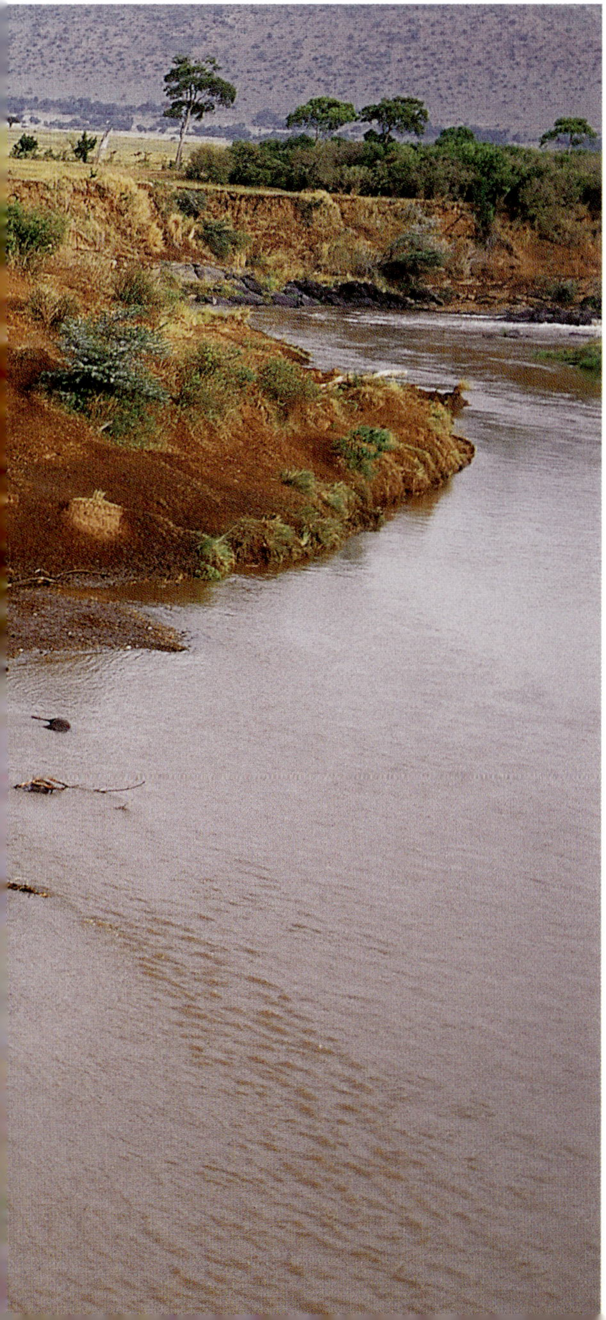

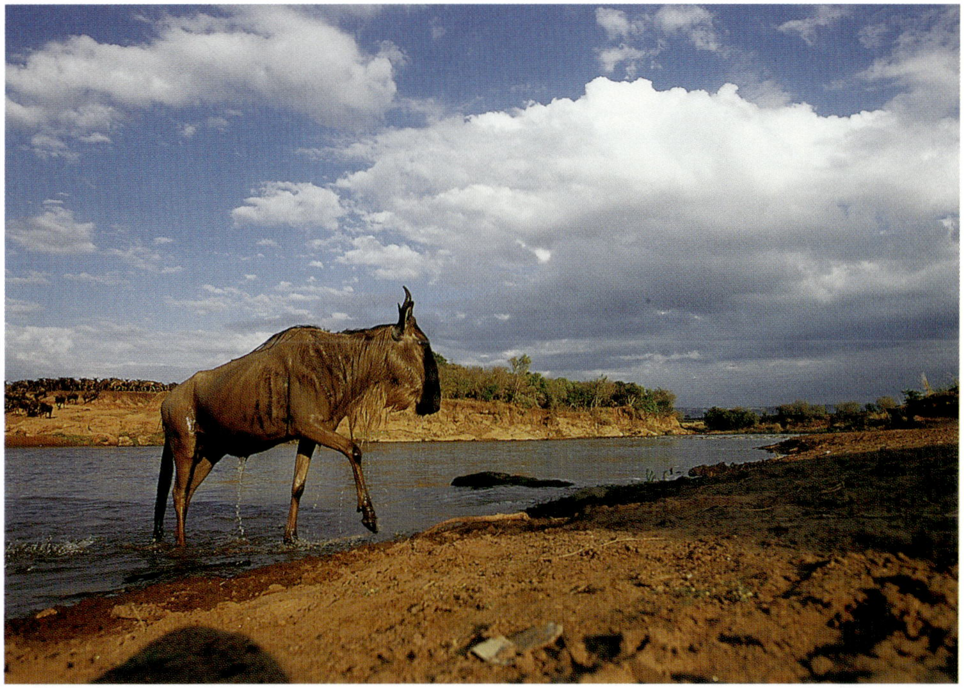

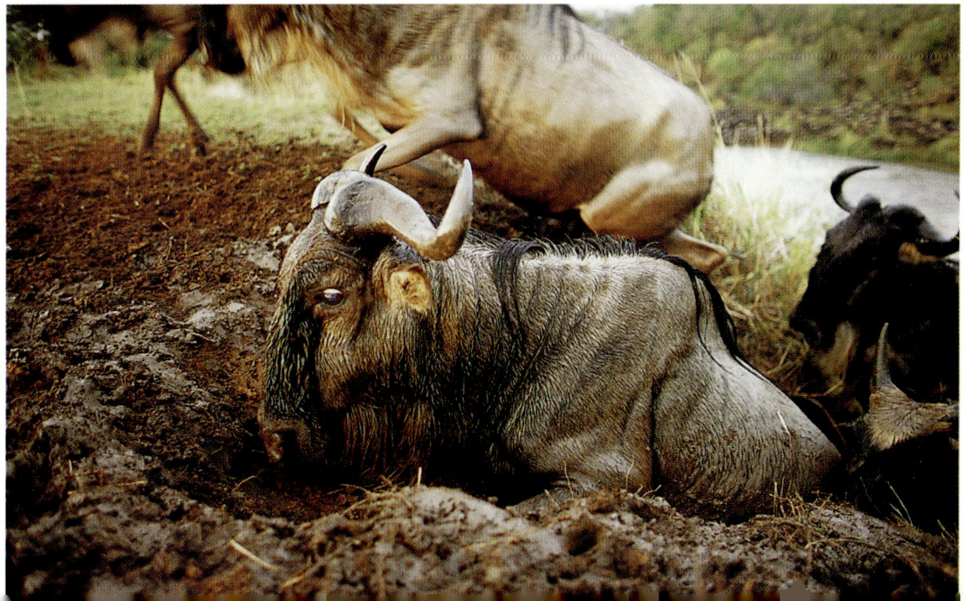

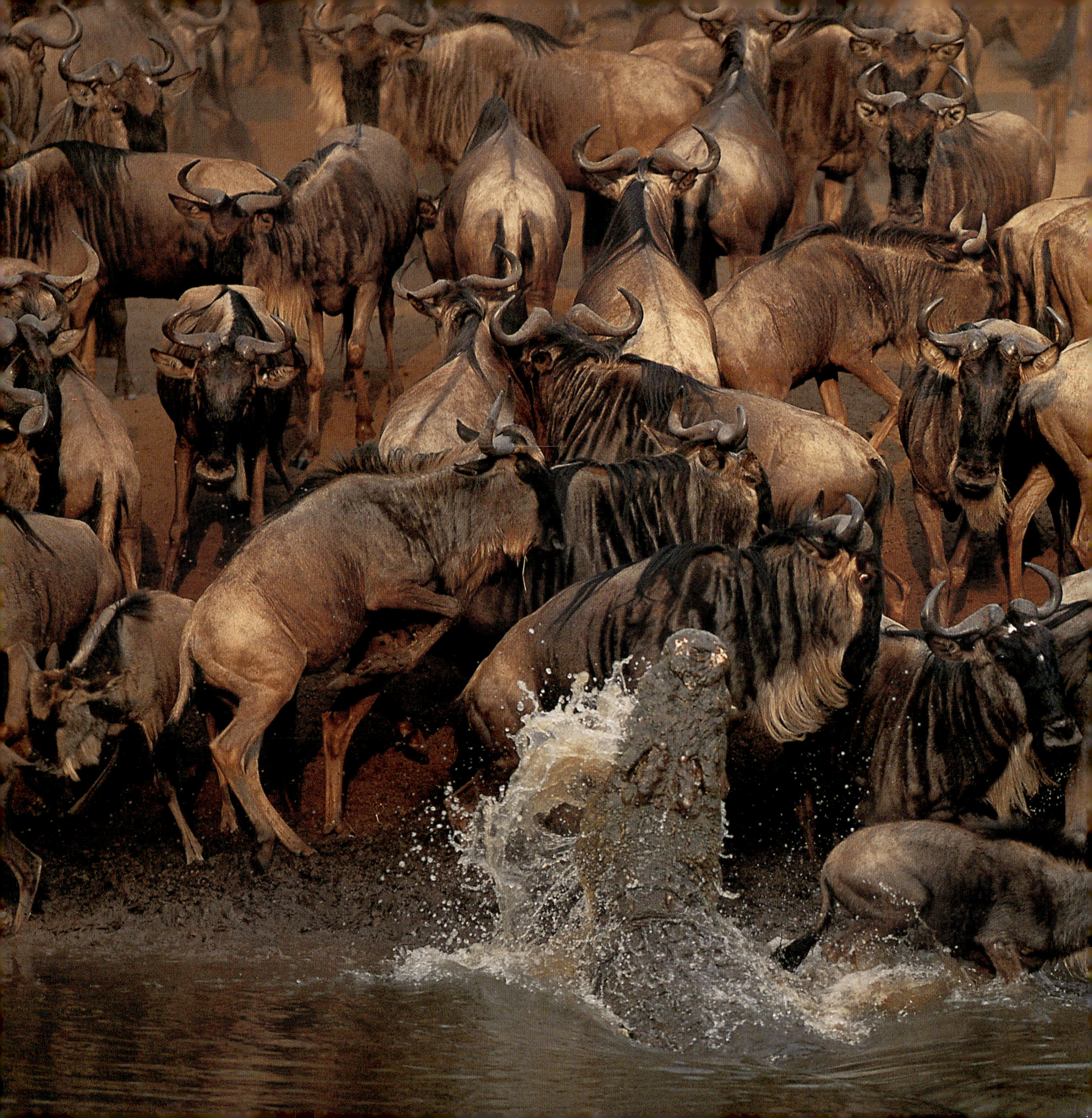

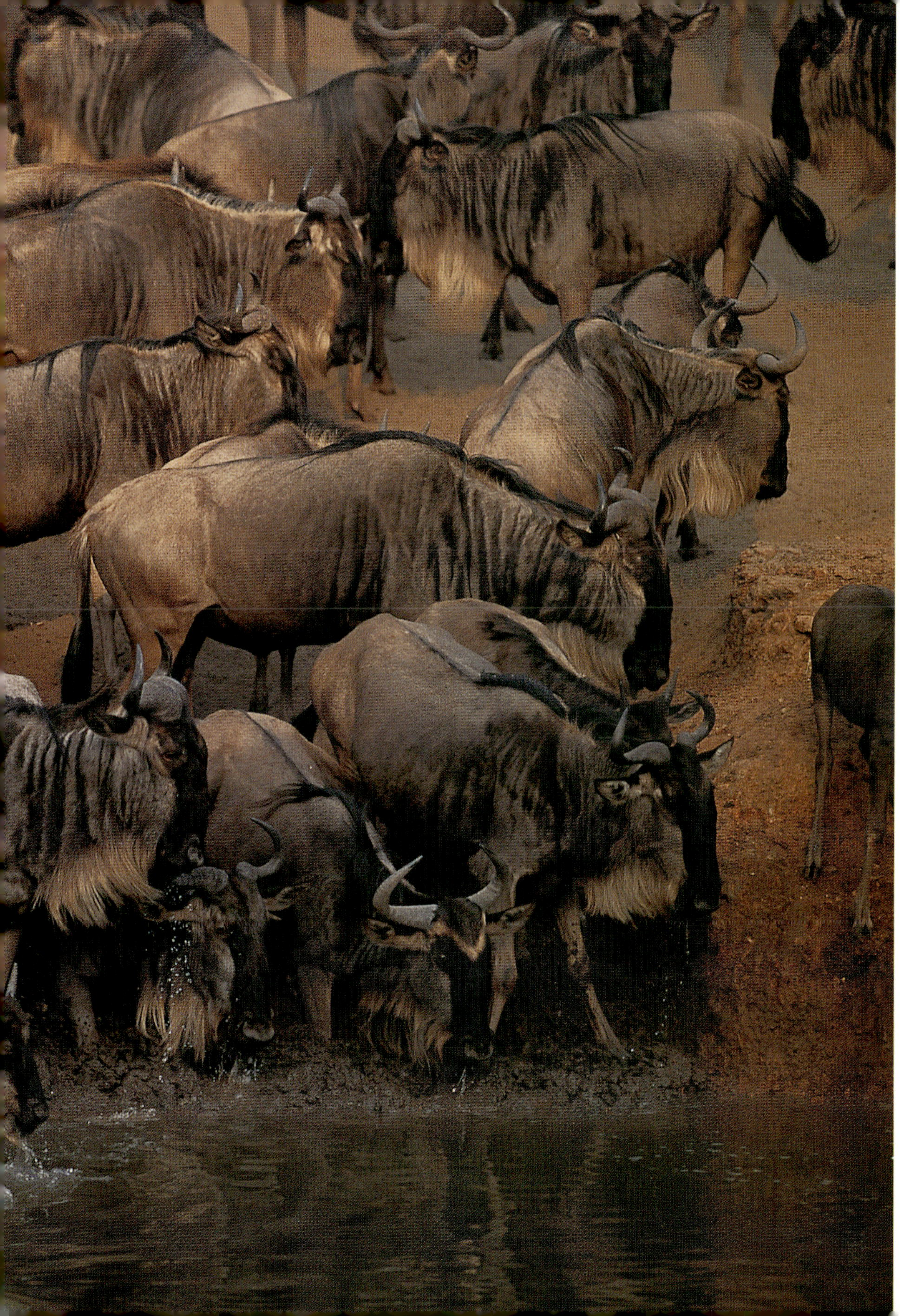

Nile crocodiles, Crocodylus niloticus, *regularly hunt drinking wildebeest in the Grumeti River by swimming underwater and emerging with an upward leap to catch their unsuspecting prey.*

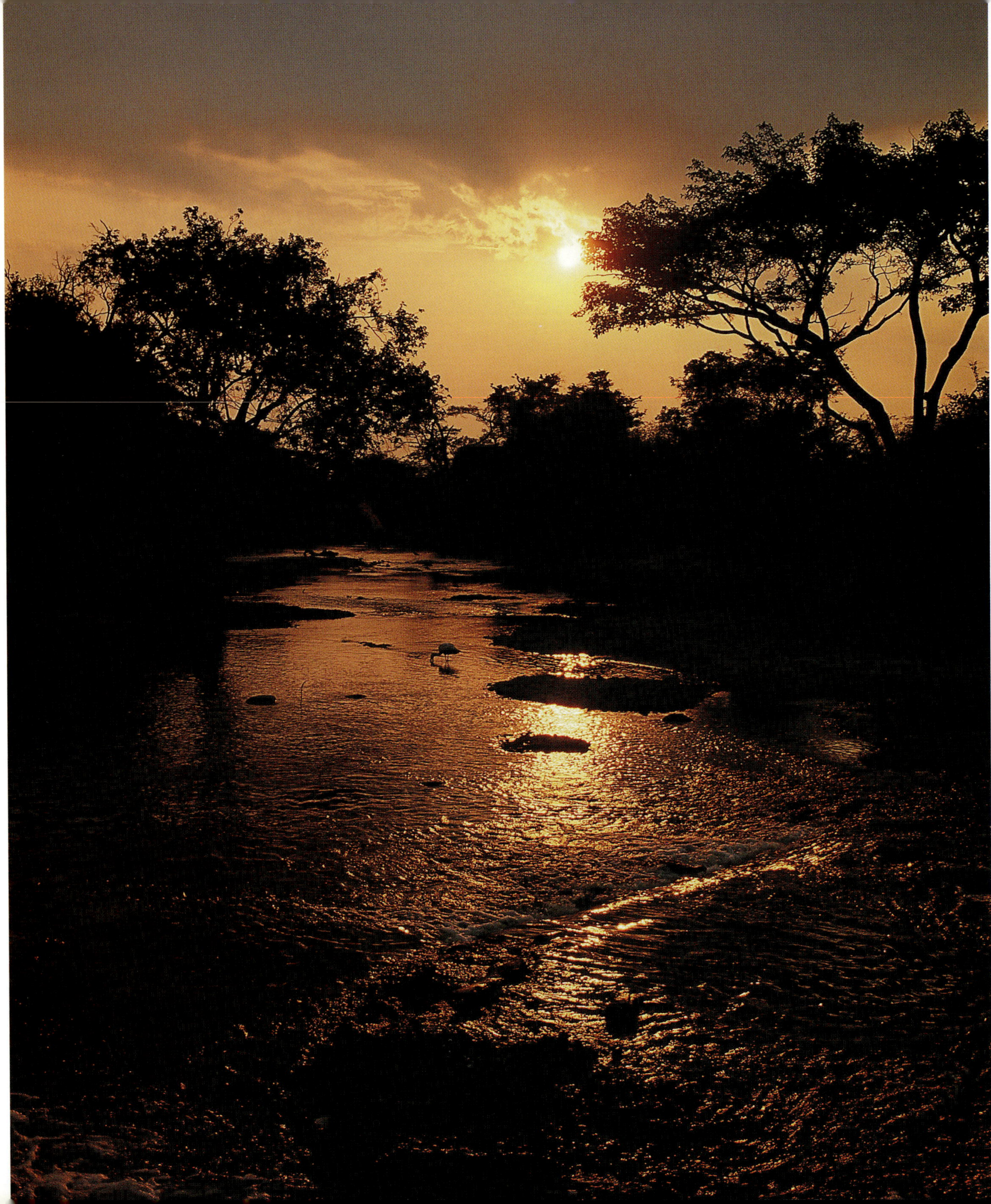

Early bipedal inhabitants

Back on the Serengeti short grass plains, at a place called Sametu, shiny black flakes of obsidian – a vitreous acidic volcanic rock resembling bottle-glass – and coarse fragments of white quartzite litter the bare ground. These stones do not occur naturally in this part of the park; they were carried to the site by ancient bipedal inhabitants. There are artefacts dated at well over 100 000 years old. The site is not very far from Olduvai Gorge and Laetoli: these plains have witnessed a prehuman and human presence for more than a million years.

The volcanic ash created conditions for grass to grow; then, from the more permanent grasslands to the north, the grazers arrived to exploit the new grass plains, followed inevitably by the carnivores. Exploiting these new resources, prehumans and humans would soon follow. The volcanoes are integral to the rift valley, and ash is their product. The rich grass plains are a consequence of all that ash: in a direct chain, both ancient and modern life are the complex and fascinating result of the rifting process.

Opposite: *The Grumeti River, flowing through the Western Corridor of the Serengeti at dusk. The river serves as a lifeline for thirsty migratory herds.*
Below: *Wildebeest calves are easier prey than adults; a young lion has ambushed and captured two calves at the water's edge.*

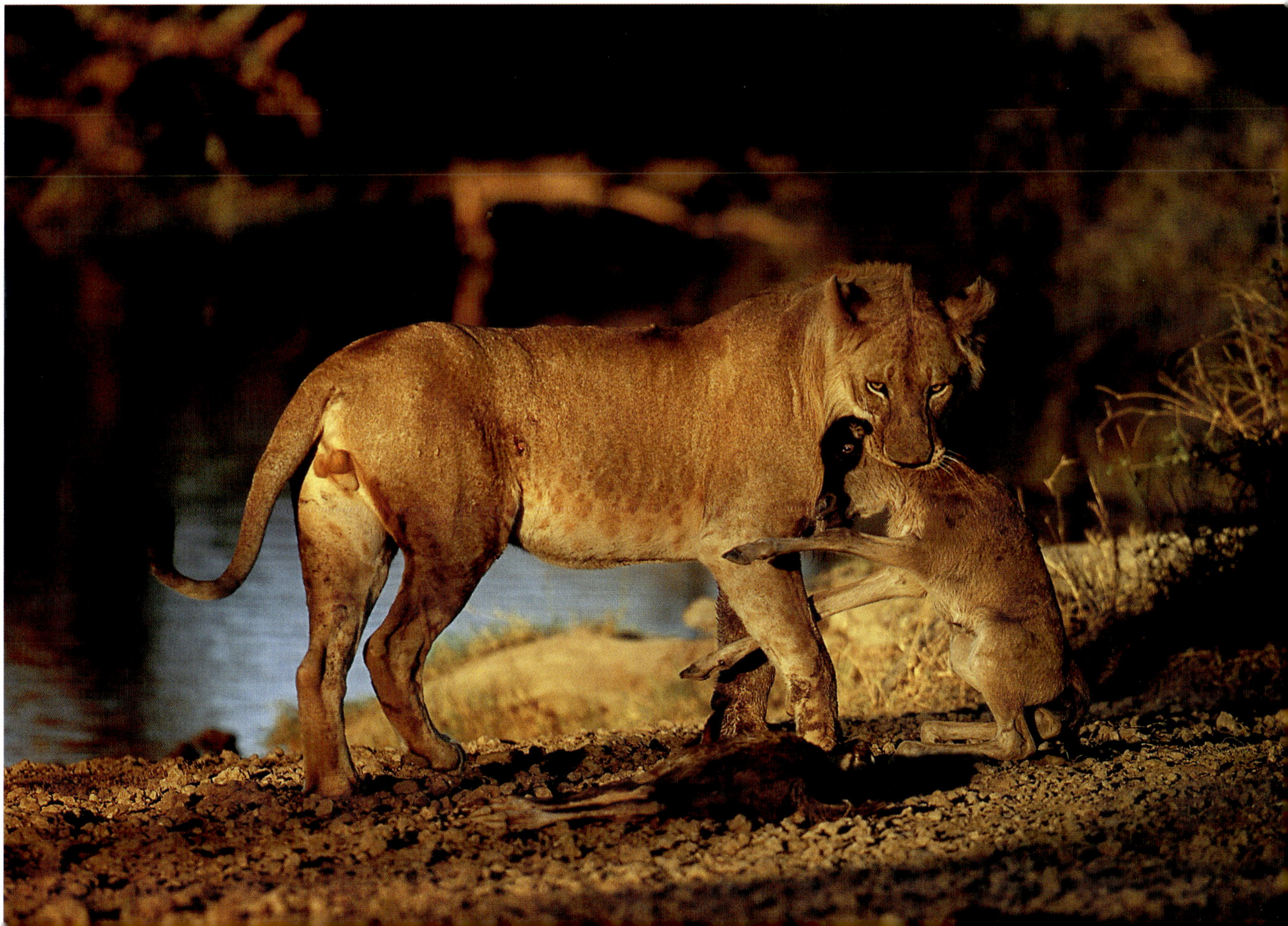

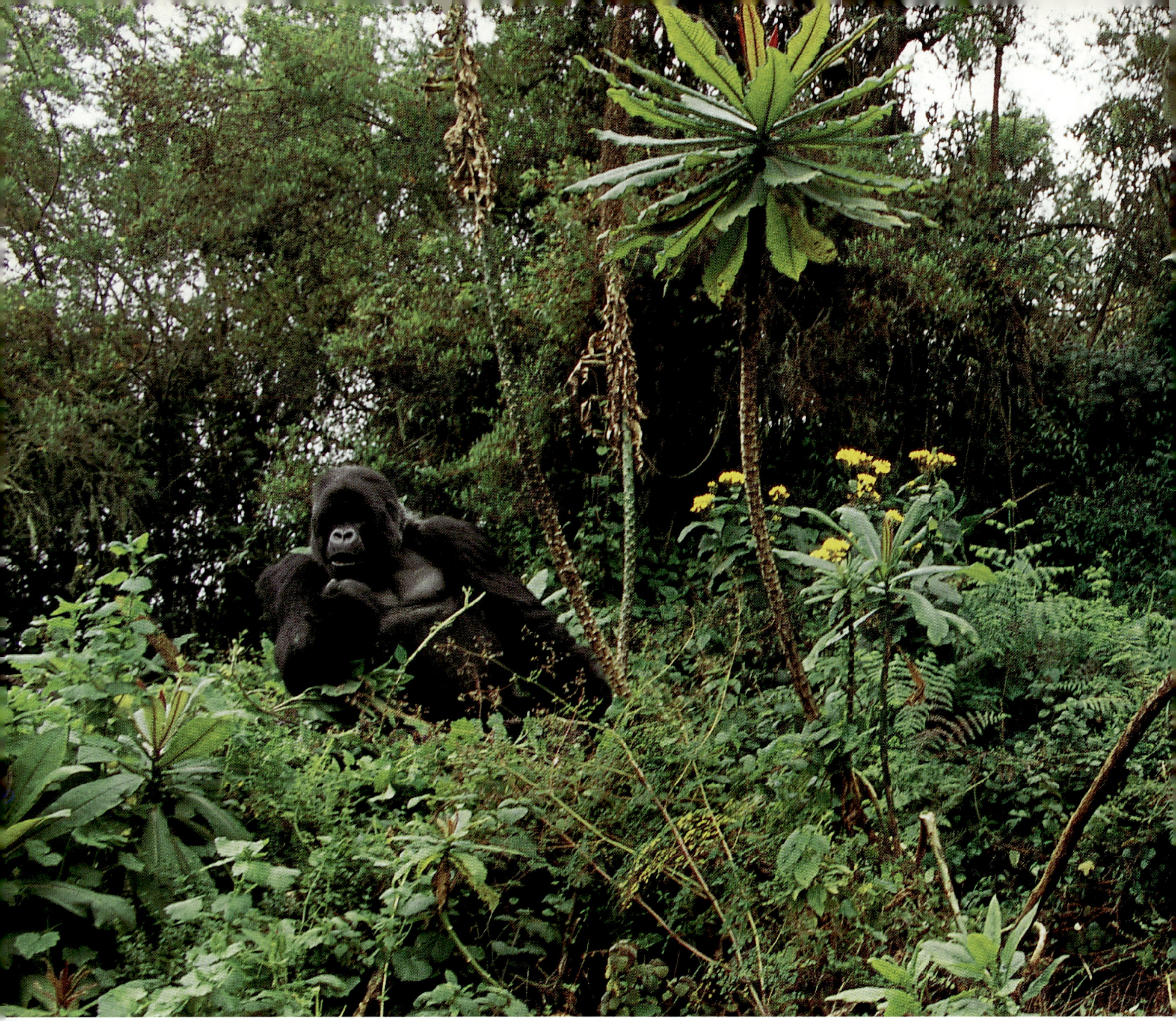

A silverback gorilla, Gorilla gorilla beringei, *in a forest on the slopes of Karisimbi Volcano, feeding quietly. This is a primaeval forest of marvellous beauty.*

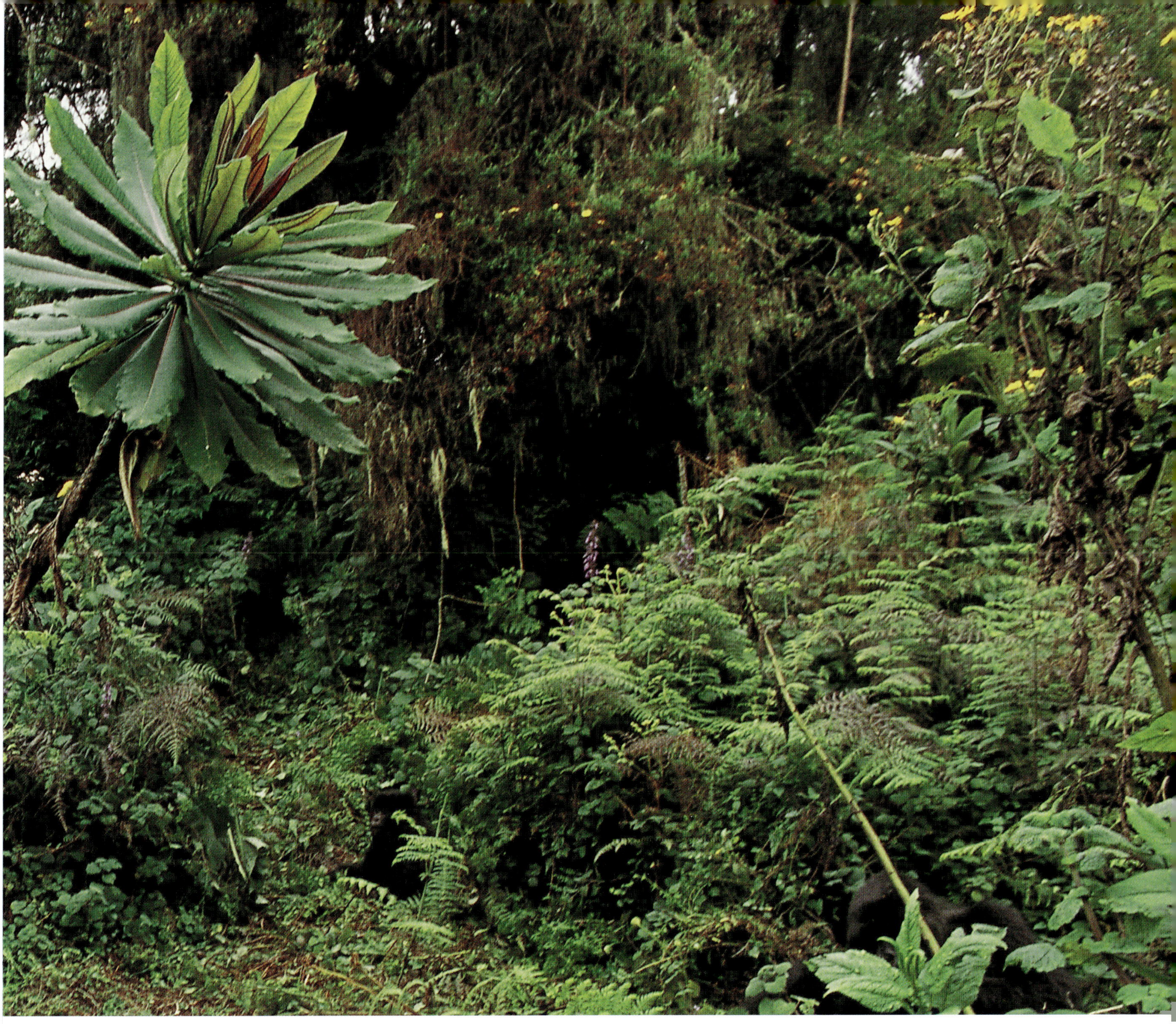

5 Gorillas of the Virungas

Gorilla evolution

Current thinking suggests that gorillas originated in the lowland rainforests of western central Africa. As the population grew, some gorillas migrated eastward, following the course of the great Congo River, but remaining on the northern bank. At one time, thousands of years ago, a population of gorillas stretched continuously from Africa's Atlantic coast to the eastern Congo. When climate change created colder and drier conditions, splitting the rainforest into two separate blocks, the eastern gorillas became isolated from the main gorilla population in the west.

At some point, the eastern gorillas reached the highlands rising along that portion of the western rift dominated by lakes Kivu, Edward and Tanganyika. The Virunga Volcanoes would have been in sight to the east, yet to reach their richly forested slopes, the gorillas would first have had to cross the extensive lava fields of the active Nyiragongo and Nyamulagira volcanoes blocking their passage from the west. At least one gorilla band must have survived the crossing and established a breeding population in the east. This group was the founder of the mountain gorilla population that we know today. Those that remained to the west of the Great Lakes became the subspecies known as Grauer's, or eastern lowland, gorillas.

Below: *Two subadult gorillas of unequal age enjoy play-fighting, while, out of sight, the remainder of the group enjoy a rest.*
Opposite: *A gorilla mother peers through the leaves of a tree.*

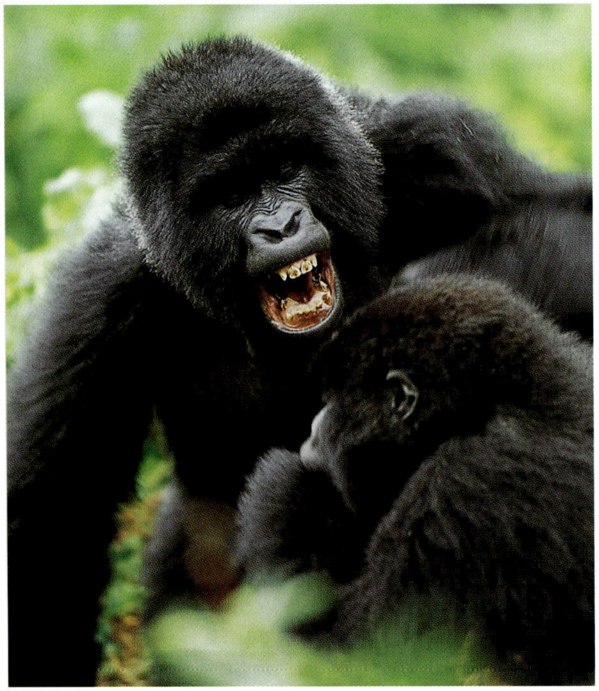

Mountain gorillas

In 1899, Oscar von Beringe, a young military officer assigned to oversee Germany's investments in Rwanda, explored Visoke Volcano and in 1902 shot two mountain gorillas on Sabinyo Volcano; in this way the Western world 'discovered' the existence of mountain gorillas. Nearly 20 years later, in 1921, the American naturalist, Carl Akeley, came to the Virungas to join a shooting party on a gorilla-hunting binge. In an about-turn, born of concern for the fate of the remaining gorillas, Akeley became a tireless crusader for their complete protection. In 1925 his efforts were rewarded: the Belgian Government granted protection to a central portion of the Virunga chain, taking in the Mikeno Volcano in the then Belgian Congo. Known as Parc National Albert, this national park – Africa's first – was in 1929 expanded to include the two active volcanoes in the west, along with gorilla habitats elsewhere on the Virungas. Today's Parc National des Virungas is the culmination of what began as a plan 'to make the world a safer place for mountain gorillas'.

George Schaller's gorilla research

Thirty years on, George Schaller arrived to study mountain gorillas during an 18-month period spanning 1959 and 1960. He was the first to follow and observe wild gorillas on an extended basis. His published observations not only laid the foundation for all future gorilla studies; they also set the standard for an entire generation of modern field biologists. Schaller carried out a census – based on direct counts and extrapolations from habitat characteristics – which produced a total population estimate of between 400 and 500 individuals.

Schaller's study yielded a wealth of information on mountain gorilla ecology. In addition to his comprehensive analysis of their behaviour and social organisation, Schaller documented the gorilla's use of diverse habitats on Mikeno Volcano in the Congolese sector of the Virungas. His direct observations produced a record of the foods consumed by the gorillas. This was a great improvement on the indirect methods of dung and stomach-content analysis that until then had been considered the only feasible way of obtaining information about wild animal diets.

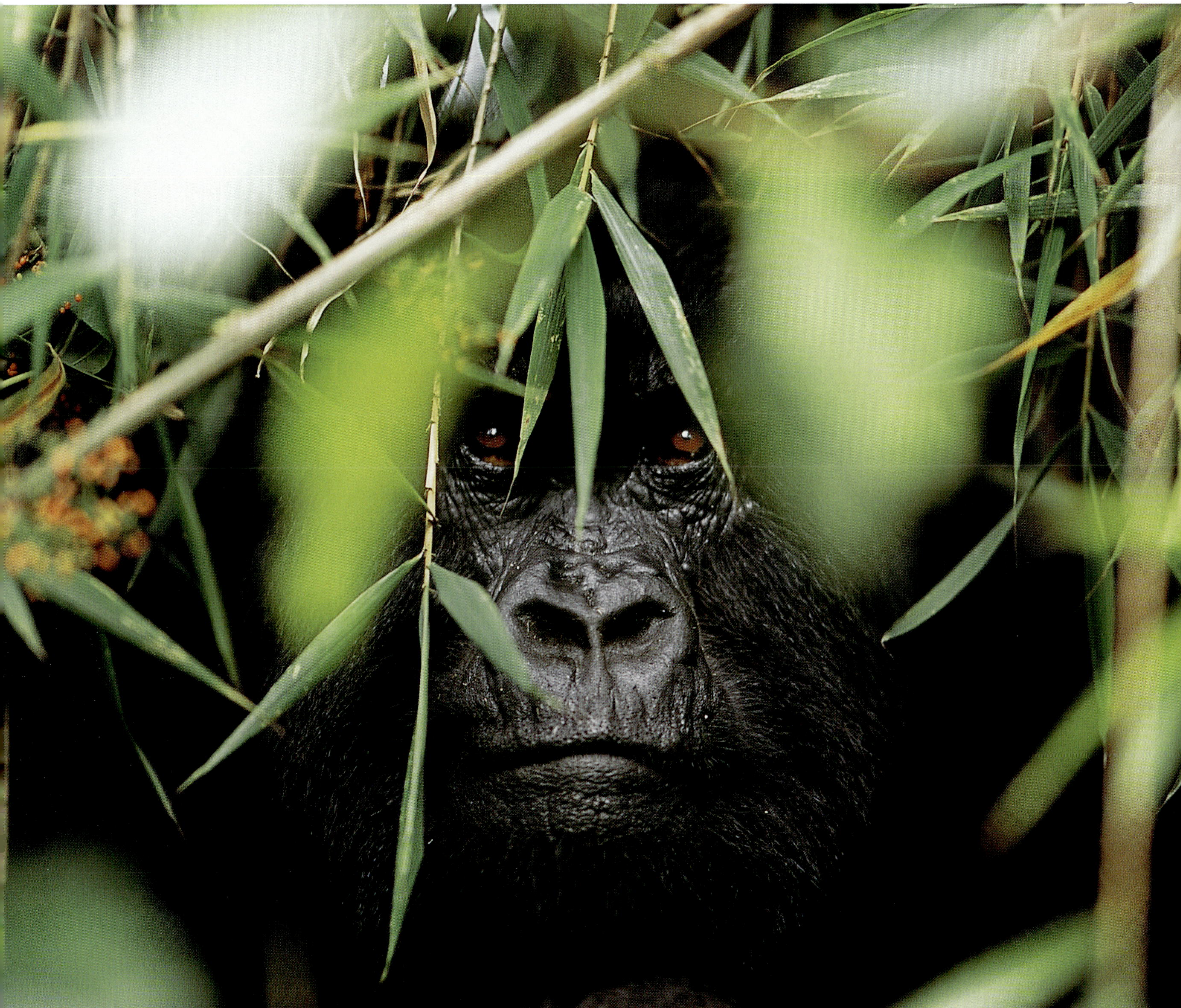

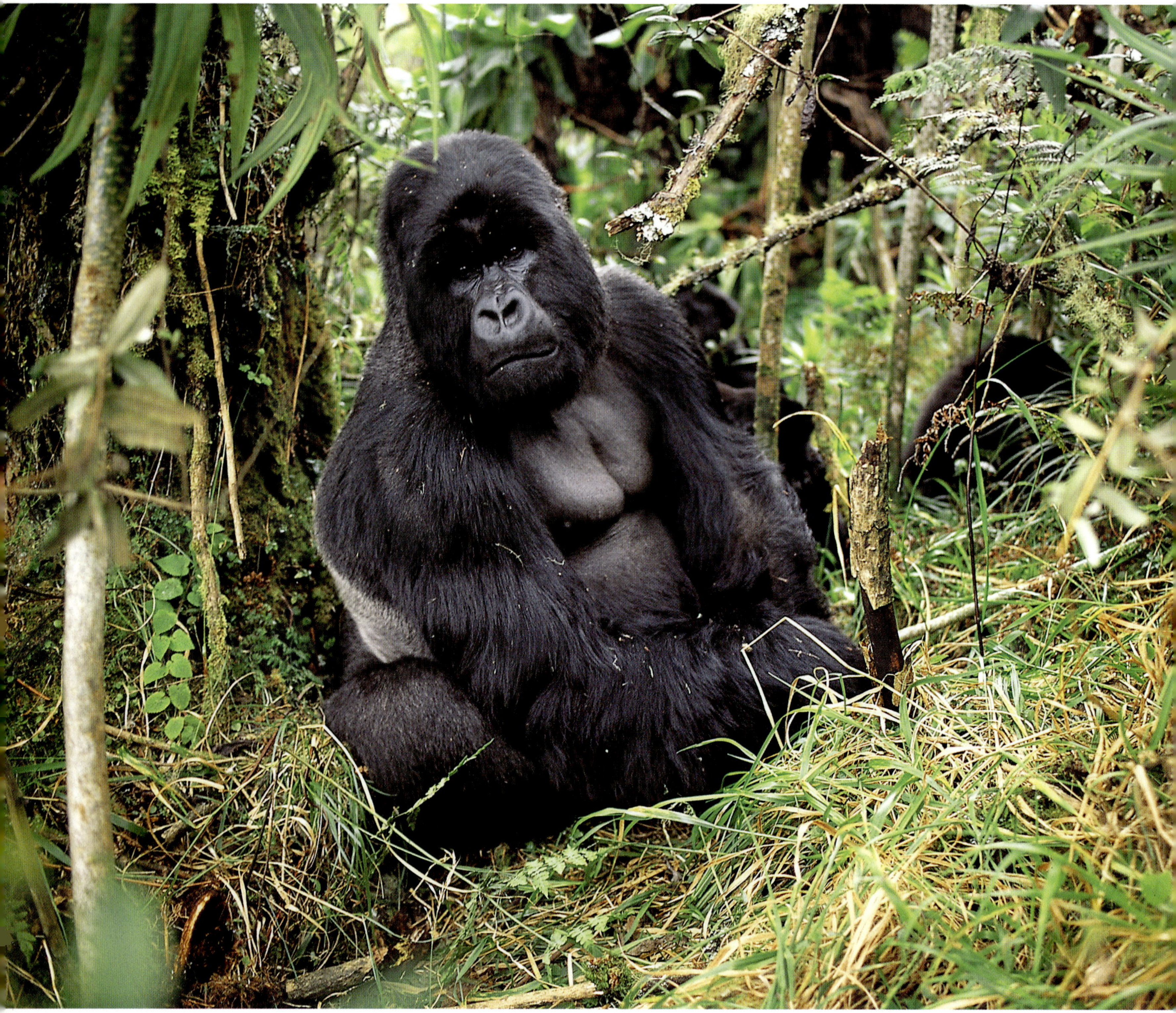

Schaller found that the Virunga gorillas make highly selective choices from among more than 100 distinctive food types. They choose more nutritious and higher quality foods wherever possible, and actively seek diversity in their food choices, although bamboo shoots dominate their diet for four months of the year. With the notable exception of bamboo, many of their food plants are widespread and abundant. Some areas with high food value are rarely visited, which suggests that the gorillas find enough sustenance elsewhere. As they feed, the gorilla families move in regular circuits, like short-distance nomads, allowing recently-browsed vegetation to recover. Gorillas also have a predilection for ants, and have been observed to engage in fights over vernonia galls infested with insect larvae. While this is far from the carnivorous (even cannibalistic) tendencies discovered in chimpanzees, it nevertheless indicates that gorillas are not pure vegetarians.

Dian Fossey's long-term study

Mindful of the contribution Jane Goodall had made to the research of chimpanzees, palaeontologist Louis Leakey was interested to meet another young woman – tall, striking, dark-eyed – who nervously approached him in March 1966 in Louisville, Kentucky, after one of his lectures on early human evolution in East Africa. Dian Fossey had been seized by his call for more research on our closest primate relatives: chimpanzee, gorilla and orangutan. Impressed by her drive, determination and conviction, Leakey secured support for her research, and Dian Fossey embarked on her lifelong study of mountain gorillas, first in the Congo, and then for many years in Rwanda.

Long-term monitoring of gorillas was Dian Fossey's greatest scientific contribution. George Schaller's 18-month study had provided a remarkably accurate portrait of gorilla social organisation and basic ecology, but this was really little more than a detailed snapshot; Dian Fossey provided something closer to a full-length feature, as she conducted an in-depth, long-term study over more than a dozen years. Characters came and went. Family dynasties formed and disintegrated. Youngsters grew up and moved on – females to other families, males to an uncertain period of bachelorhood. Matrilines emerged as a powerfully cohesive force in gorilla society. Dominant females demonstrated a capacity for leadership in many situations. Patterns of succession between silverbacks slowly appeared. Birth and death rates and other demographic trends could be calculated from the expanding set of baseline data. Despite the comings and goings of individuals, group conflicts that were punctuated by spectacular displays, and several infanticides recorded over the years, the gorillas earned the reputation of having a gentle disposition despite their great size. Unlike chimpanzees, which showed a violent streak, mountain gorillas generally kept the peace. Fossey chose a different strategy from Jane Goodall (who had at first provisioned chimpanzees) in deciding not to feed the gorillas. The result was a reflection of natural feeding behaviour and a reliable information base on gorilla habitat and food requirements. Into the twenty-first century, researchers continue to monitor some of Fossey's original study gorillas and a host of their offspring.

Opposite: *Gorillas are the largest of the living apes. An adult male can reach nearly 1.8 metres (six feet) in height and weigh180 kilograms (400 pounds). Adult females weigh in at around 90 kilograms (200 pounds). Their size belies a largely gentle disposition.*

The gorilla harem

The gorilla's social system, evolving over many years, has been shaped by the differing needs of females and males in their long-term struggle to survive and reproduce. Females, constrained by the high costs of reproduction and parental care, place the highest premium on ensuring access to sufficient energy, in the direct form of food. For males it is very different – the costs of reproduction are low, and they are driven by access to females, rather than food. While

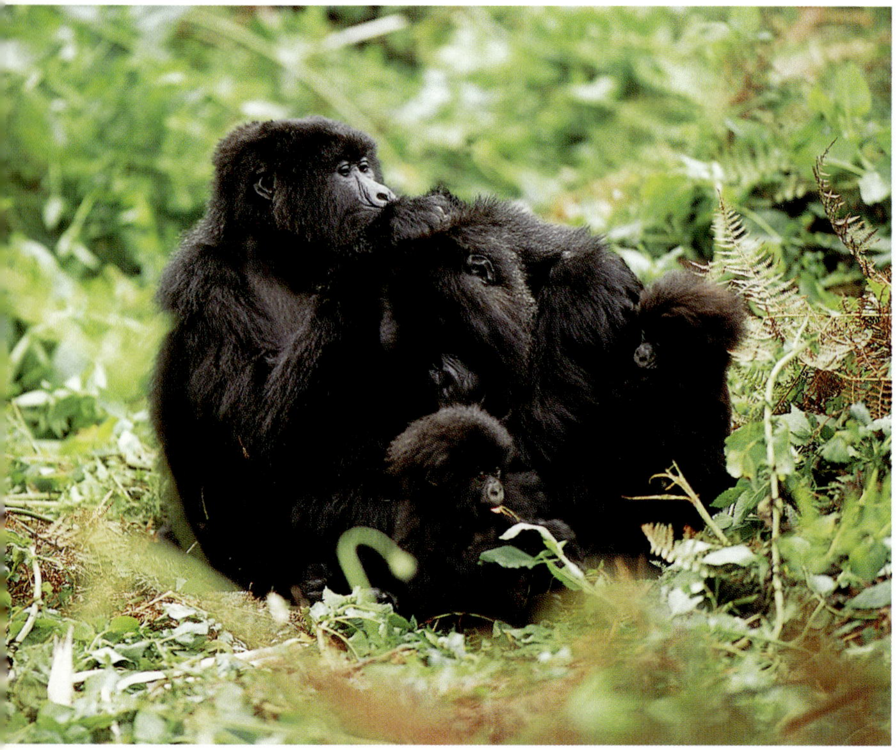

females distribute themselves around the food sources, males distribute themselves around the females. For mountain gorillas, food resources, located in large, uniform patches on the Virunga slopes, are abundant. As a result, females are able to survive and reproduce within very small home ranges and, in turn, single males are able to defend groups of females, or harems. The harem system, adopted by gorillas, is not, however, a dominant feature of human social heritage – one of many features of gorilla behaviour suggesting that prehumans were quite different from gorillas.

Large individuals tend to live a long time. At 12 years of age, a male gorilla starts to develop silver hair on his back, and at 14 years, his back is fully silver. An established silverback is likely to live for over 40 years. Dominance takes time to establish, but 15 years of dominance is not unusual at the head of a harem. By the time he dies, the dominant silverback should have produced a successor from within his own bloodline. Only one male will inherit his father's harem; the rest will disperse in the hope of starting independent harems.

When females reach sexual maturity, at around eight years, they normally transfer to other families, or meet up with a wandering male seeking to start a harem. Hierarchies are enforced not only within groups, but also between lineages, where vocal hostilities may arise over access to choice foods. The hostilities can be intense – opponents often scream and posture – but the outcome predictably favours the dominant matrilineage. In rare instances, the silverback may intervene.

Above: *Rest time is also grooming time. Two mothers, each with an infant, engage in mutual grooming.*
Opposite: *A youngster munches on stems and leaves. Mountain gorillas live in a rich 'salad bowl' of diverse edible plants, and consume as much as 23 kilograms (50 pounds) of leaves, stems and roots in a day.*

Silvered across much of his body, and weighing around 180 kilograms (400 pounds), the silverback is both an imposing figure and a calming presence. Social skills are an important part of gorilla leadership; the dominant male reinforces his authority through a network of relationships within the group.

Mountain gorillas best reveal their social nature during the late morning or midday rest period – siesta time for some, but grooming time for others. Once on the move, however, gorillas operate in a competitive context. Food resources, while generally abundant, may be limited in both quantity and quality at any one place or time. Under circumstances of conflict, hierarchical status is usually invoked to resolve any vocal disputes over delicacies.

Early development

Gorilla infants, weighing between 1.4 and 2 kilograms (3 and 4.2 pounds) at birth, spend their first three years in close physical contact with their mothers, who nurse them and with whom they sleep. During the first six months, infants depend almost entirely on rich mother's milk. They are extensively and frequently groomed. Almost never out of direct maternal contact, they are held in their mother's arms during feeding, lie on their mother's stomach during rest, and hold on to their chest or back as they move through thick vegetation. Clinging to their mothers' hair, they not only see what their mothers are eating, but are also showered with a steady rain of edible remains. Mouthing these morsels without swallowing gives them a first taste of what lies beyond mother's milk.

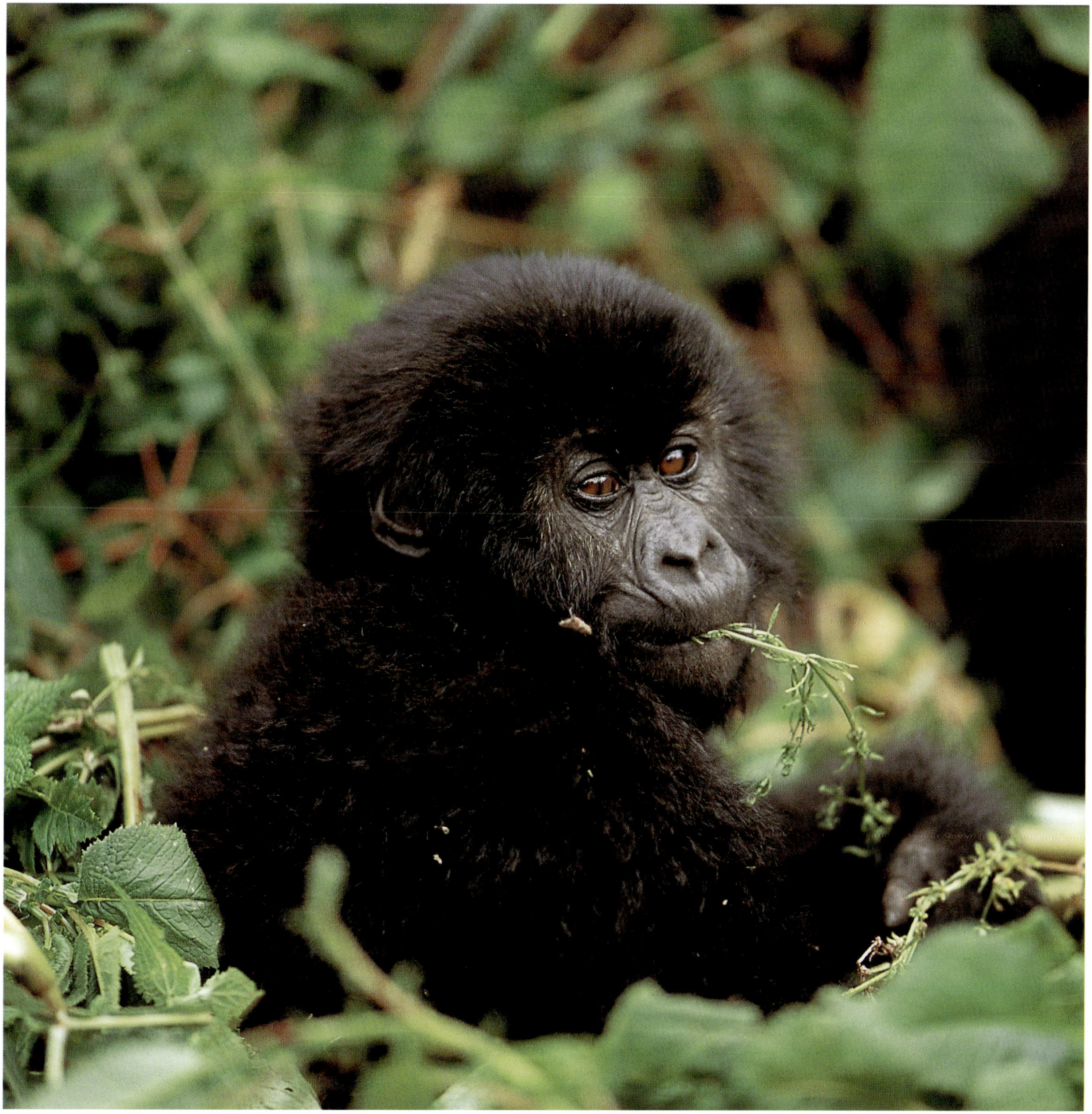

By the age of six months, most infants begin to experiment with solid foods. Gorilla mothers 'instruct' their young in what is edible and – more importantly, in a world with many poisonous plants – what is not, by confiscating inappropriate items. For the next year, infants are given some limited freedom of movement, but stay for the most part within their mothers' reach or sight. Even at rest, they are often constrained by a discreet, yet forceful, maternal footlock around the ankle or wrist. Their roaming distance increases steadily through the third year, as their interactions with other infants and older juveniles multiply dramatically. At this stage, juvenile gorillas imitate the eating habits of those around them. This is also the age at which individual or personality differences start to appear.

Gorilla games are similar to those found in most human cultures around the world. Tag, wrestling and 'king-of-the-castle' all have their Virunga variations. Play is most common during group rest periods, when two or more young gorillas chase each other through the surrounding underbrush and then engage in a freestyle wrestling match. They play for as much as one-third of their waking hours, sometimes even enticing their elders to join in. Gorilla youngsters may not laugh or scream, but they do emit a stuttering 'chuckle', and apparently enjoy the games enormously. Gorilla childhood – a long period of dependence on the mother, with a high premium placed on learning, and a great deal of time devoted to play – is reminiscent of human childhood.

Not our closest relatives

In several important ways, gorillas differ from chimpanzees and early humans: whereas chimpanzees have been observed fashioning and using tools, there are currently few such observations among wild gorillas. Gorillas do not appear to practise deception; they also perform poorly in other cognition tests such as the 'mirror mark test' for self-recognition, and tests of imitation. The cognition of gorillas is an issue that stimulates heated debate. Since mountain gorillas do not hunt, the occasion for co-operative hunting activities has not arisen; the extent to which co-operation exists in other spheres of gorilla activity is not well researched.

Unlike their human ancestors, and, to a certain extent, unlike chimpanzees, gorillas stayed on in the forests. Today, mountain gorillas live on isolated forest 'islands'. Although they do venture up the slopes to forage in higher pastures, they are effectively cut off further down the slopes by advancing farmland. Despite their isolation, their home is bountiful and comfortable. They live a life of relative ease, consuming nothing more than enormous quantities of leaves, stems and bamboo during peaceful feeding forays around their abundantly stocked forest domain. The highly endangered mountain gorillas may have been stranded on the Virunga mountain slopes, but they have evolved in such harmony with their surroundings that, were it not for the threats they now face, they might appear to be living in a gorilla paradise.

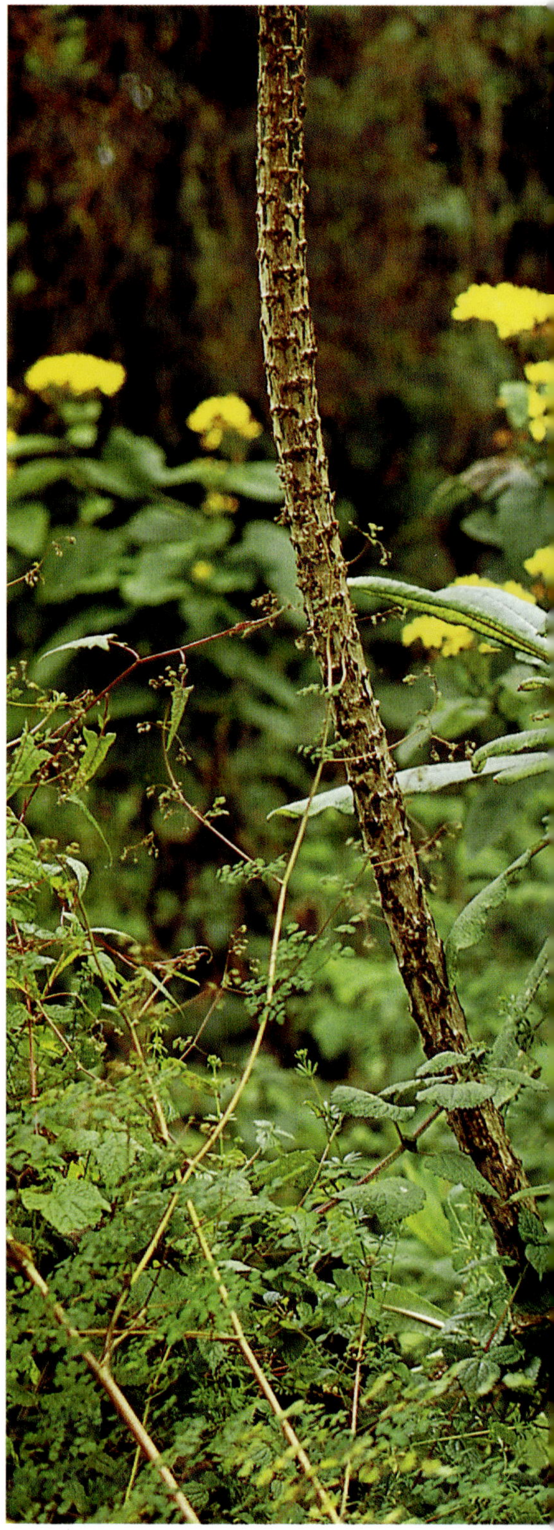

Play, both alone and socially, is a central element in a young gorilla's life. This youngster appears to be climbing purely for fun.

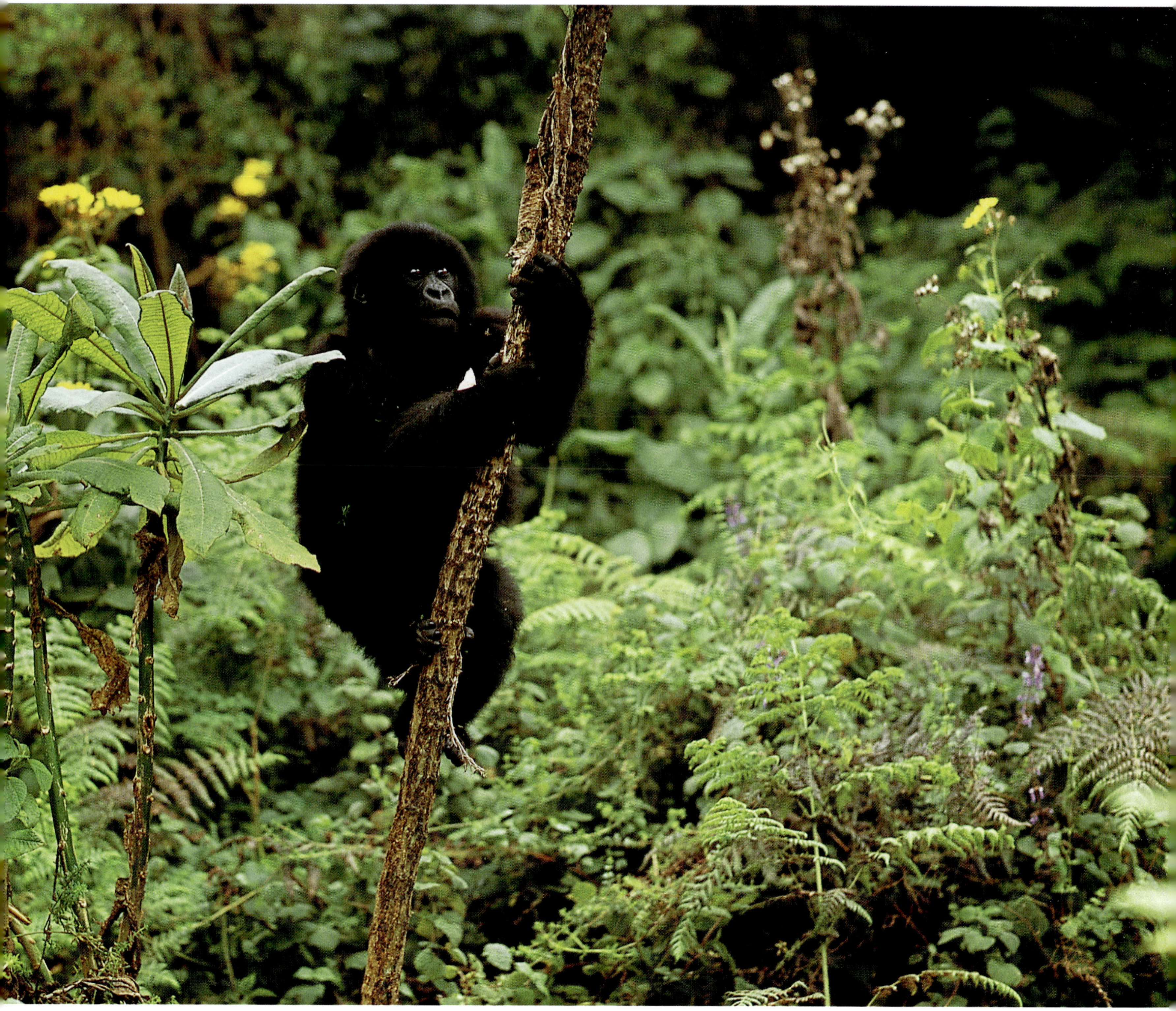

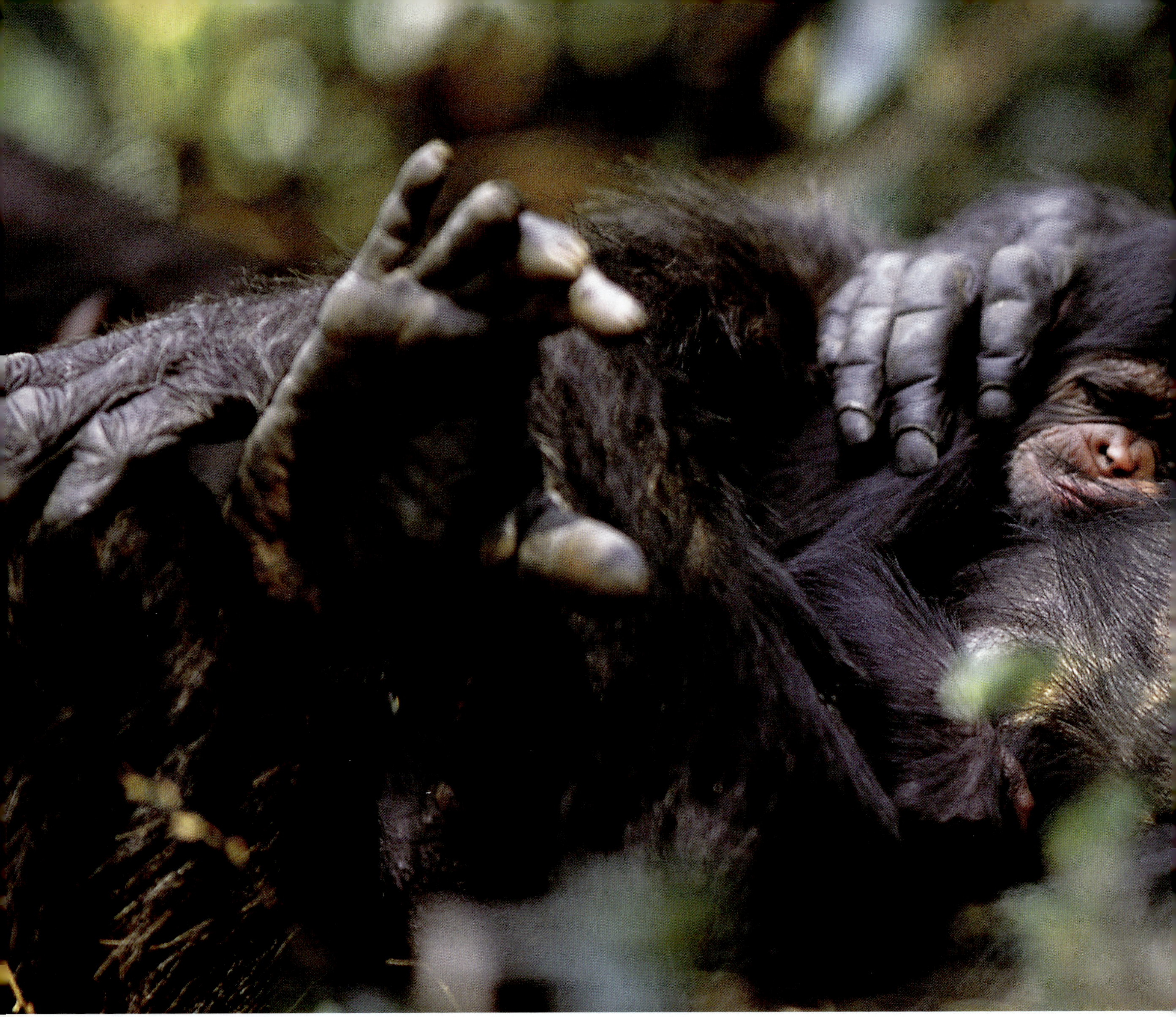

During her lifespan of over 45 years, Fifi, a chimpanzee, Pan troglodytes schweinfurtheii, *of Gombe, bore eight young, including two alpha males. This image shows her cradling two-day-old Furaha.*

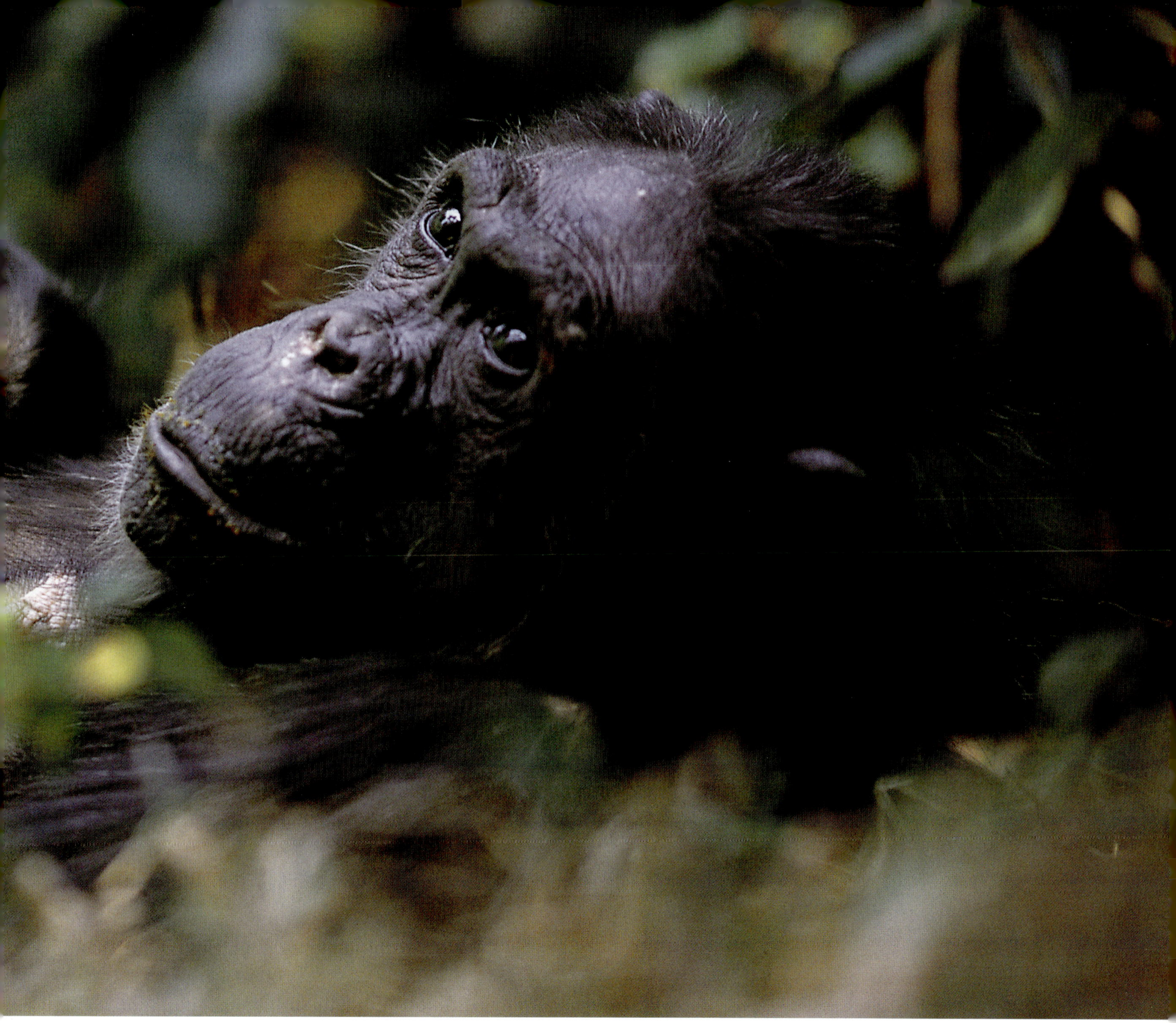

6 Chimpanzees of Lake Tanganyika

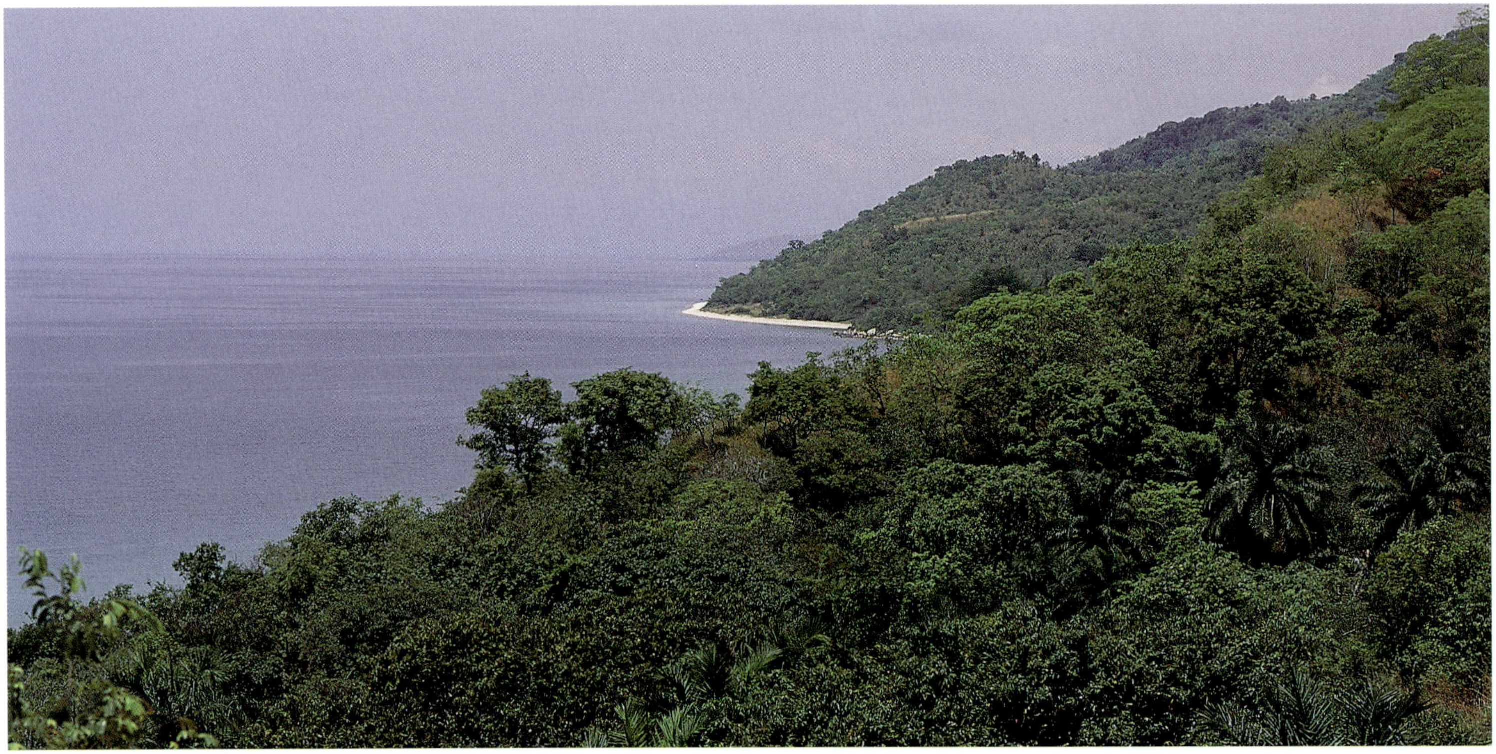

Although Gombe National Park, on the shores of Lake Tanganyika, receives a lot of rain, there is a marked dry season between June and October, with the result that the forest is mainly 'open canopy' – it has mixed vegetation and clearings where breaks in the foliage allow in light.

Both Gombe Stream National Park and Mahale Mountains National Park, 170 kilometres (105 miles) apart, are gallery forests arising from the eastern shores of Lake Tanganyika. These forests are home to populations of chimpanzees, and the primates have not only become the focus of much public interest, but have also drawn scientists to Africa's Great Lakes region in the search for the origins of man.

Lake Tanganyika is a lengthy section of the southern portion of the western rift that has simply filled up with water. Its walls are the walls of the rift. Since the lake, which is 1 436 metres (4 710 feet) deep, lies 767 metres (2 516 feet) above sea level, the floor of the rift valley on the lake bottom is 669 metres (2 194 feet) below sea level, making this the deepest area in the whole of the rift valley. The lake temperature remains surprisingly constant from surface to lake bed, with a gradient of only 3 °C. Because the water does not circulate, the deeper layers of the lake are permanently undisturbed and deoxygenated, and thus biologically inert. No life exists 198 metres (650 feet) below the surface. The lake's crystal-clear upper layers, by contrast, are oxygenated, and support abundant life. The proliferation of cichlid fish species that have evolved here – and nowhere else – emphasises the long period over which the lake has existed in isolation.

Lake Tanganyika has over its long history – it is the oldest of Africa's great lakes – endured several periods of dramatic climate change, all of which have shaped its unique aquatic ecology. Today, as water temperatures continue to rise, and as inflowing rivers bring more and more sediments into the lake from an increasingly deforested and eroded catchment area, that adaptive ecology is again undergoing a process of rapid transformation. How this will affect the many cichlids and other organisms that inhabit the lake's clear, sparkling waters, and the chimpanzees that have survived in its remaining isolated shoreline forests, remains to be seen.

Chimpanzees of the forests

While stationed in Kenya with the British Army in 1945, Cambridge anthropologist Jack Trevor took leave to visit a forest bordering the shores of Lake Tanganyika in Tanzania. He encountered and tracked wild chimpanzees in the forest and, on his return, shared the experience with fellow anthropologist, Louis Leakey. Trevor described the area as an open gallery forest, with rivers running down into a lake, interspersed with patches of grassland. This could have been a description of the habitat Leakey had predicted for the prehuman apes that had emerged from the thick tropical forests to forage in more open country. Leakey believed that the behaviour of these gallery forest chimpanzees could throw light on the behaviour of prehumans and, by extension, the evolution of early human behaviour.

Jane Goodall's Gombe study

Leakey was keen to pursue his investigation of human origins, and gave the task of studying the chimpanzees of Gombe to his assistant at the museum in Nairobi – Jane Goodall, a young Englishwoman passionate about working with animals. In the summer of 1960, Goodall, her mother and a cook established a camp on the shore of Lake Tanganyika in Gombe Reserve, where she was to begin a sustained study of chimpanzees.

Goodall observed aspects of chimpanzee behaviour that had never before been reported, among them the earliest documented use of tools by a chimpanzee. Like humans, chimpanzees have opposable thumbs, and can apply either a powerful grasp, using the whole hand, or a precision grip, between finger and thumb. To extract

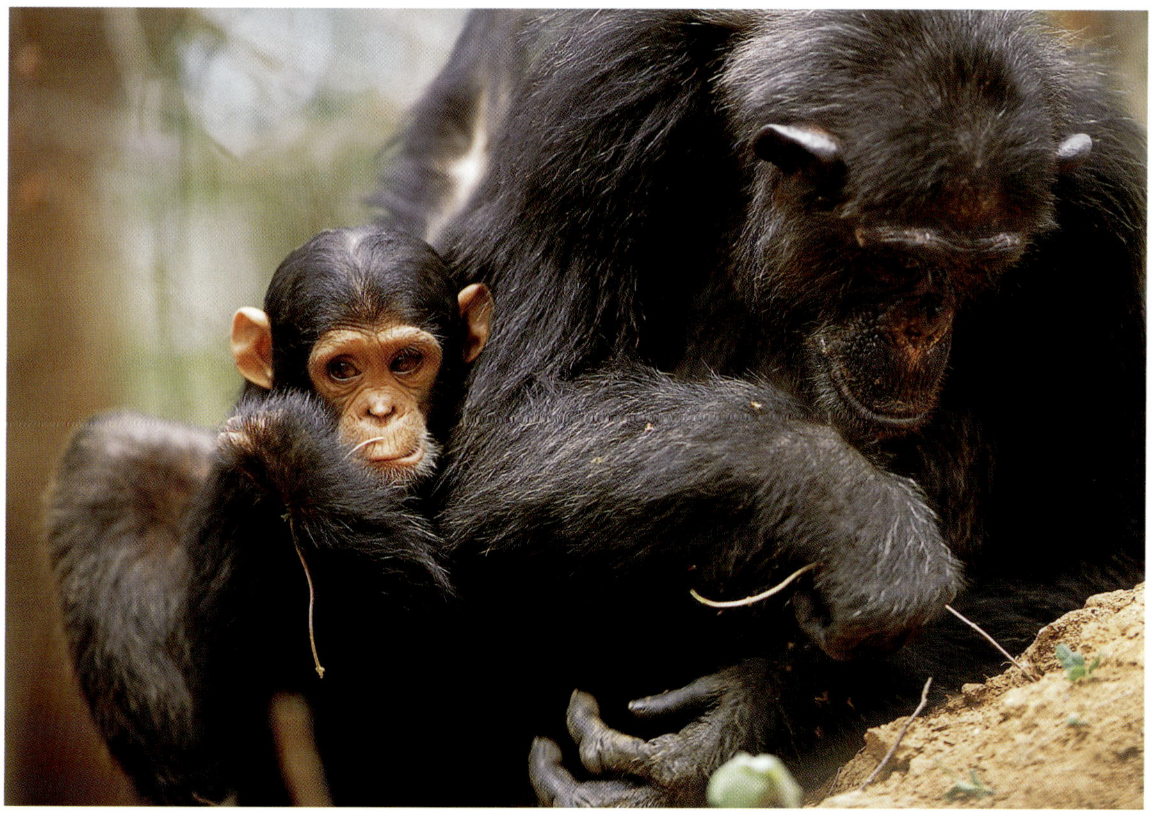

Tool use is passed down from one generation to the next. Fifi demonstrates using a plucked stem to extract soldier termites from a mound. Her son, Furaha, will over time learn this technique from his mother.

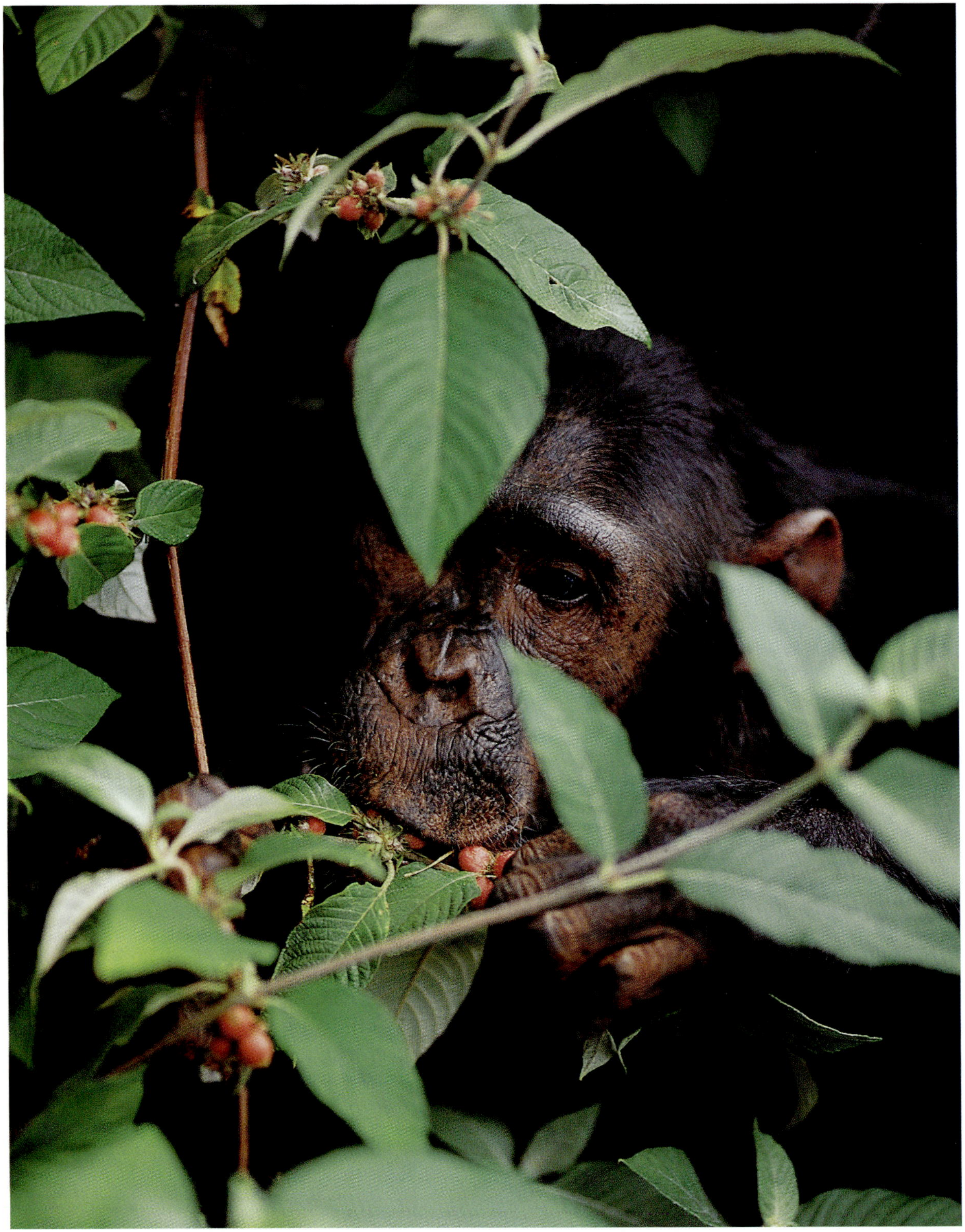

Fifi's adolescent son, Faustino, eats all the ripe berries he can reach. Fruit is the mainstay of the chimpanzee diet.

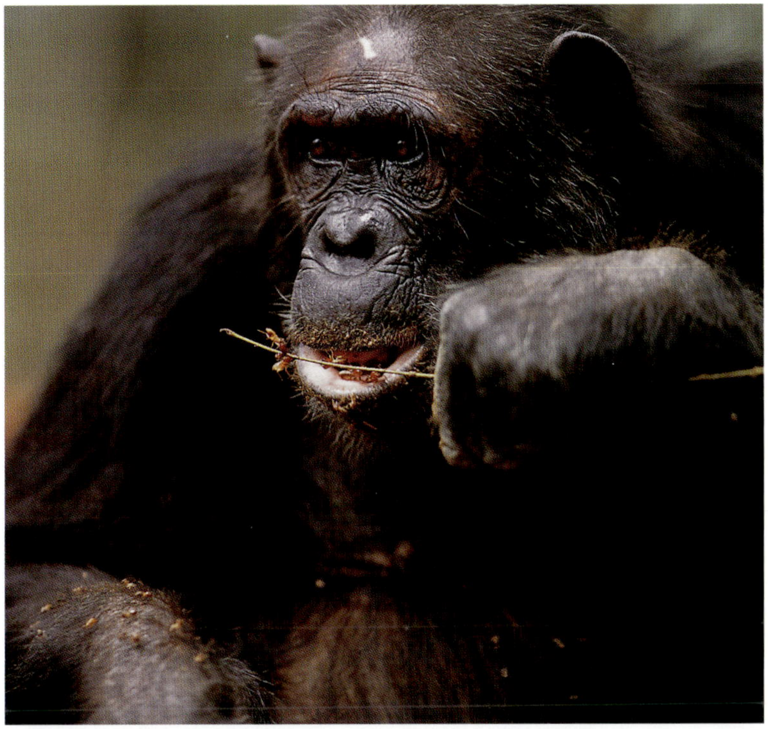

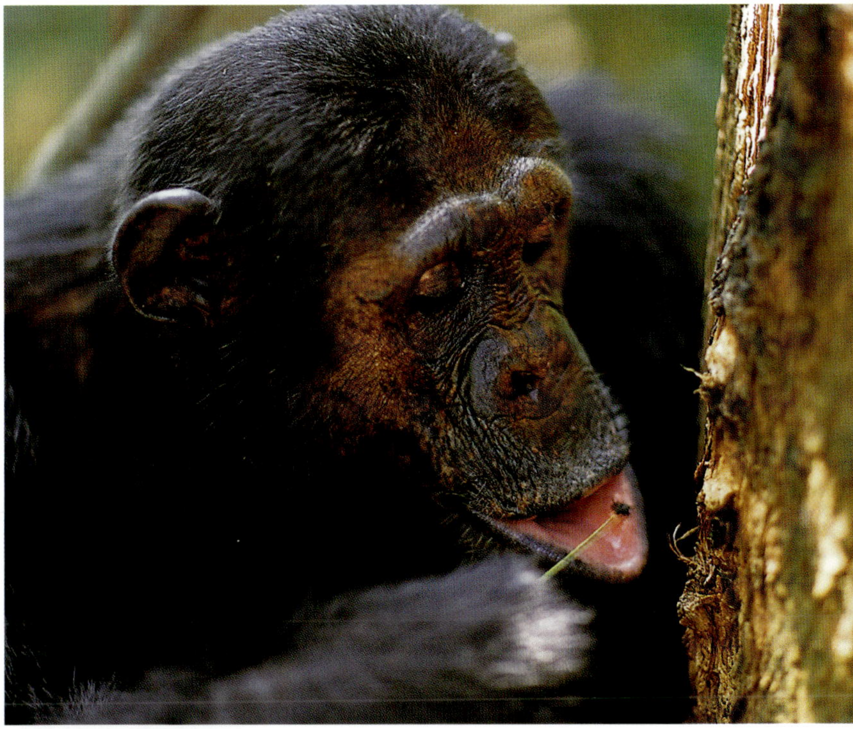

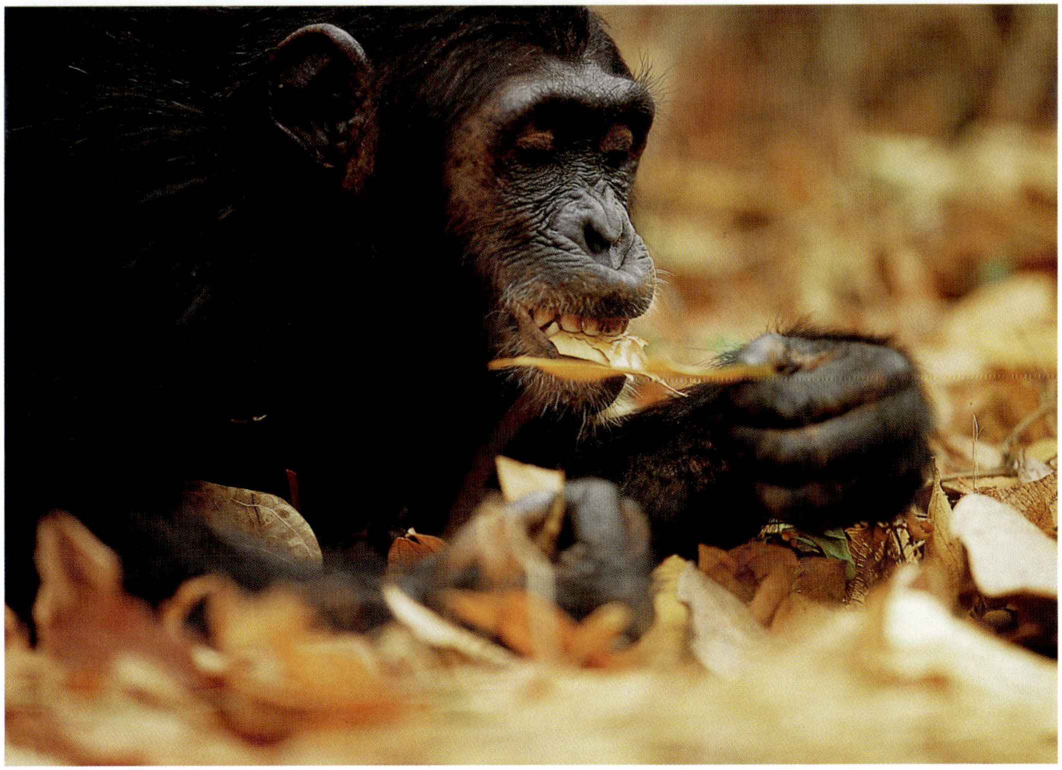

***Above left:** The tool is inserted deep into a twisting termite tunnel, then cautiously withdrawn so the termites are not dislodged. Gremlin, a female at Gombe, typically catches several hundred termites an hour.*

***Above:** Alouf, alpha male at Mahale, dips a stem stripped of leaves into a hole in a tree trunk. The defending ants, grasping it, are extracted and delicately consumed.*

***Left:** Without eating them, a chimpanzee bites little bits off stiff leaves with its front teeth. This 'leaf-clipping' produces a distinctive sound. At Mahale, young males and adult females that are ready to mate make the clipping sound to attract each other's attention.*

termites from their nests, chimpanzees strip leaves off a stick and insert the thinner end into a hole in the nest; this probe is soon covered with tenaciously clinging, agitated termites, which the chimpanzee proceeds to devour. To obtain water collected in holes in trees, chimpanzees chew leaves to create an absorbent sponge for dipping into inaccessible reservoirs. Before Goodall's discoveries, it was widely accepted that only humans used tools.

When Goodall observed an adult male eating the remains of a young bushpig, she was able to challenge the accepted model of the vegetarian chimpanzee. She went on to observe that chimpanzees were formidable, efficient and co-operative hunters, able to stalk, ambush and out-manoeuvre a variety of prey species, including agile monkeys. Subsequent studies have revealed that, although meat accounts for only five per cent of the chimpanzee diet, it is a highly prized delicacy. When a prey animal is captured, it is usually given to the alpha male, who soon finds himself surrounded by begging individuals. Apart from the portion he himself devours, he shares out the spoils according to a simple strategy: ignoring the entreaties of pretenders to his throne (young males of rising rank and beta males), he shares the food with non-threatening, middle-ranking males, with old but influential males and with persistent begging females. Others have to settle for the scraps that fall to the ground. This strategy maintains the position of the alpha male by exerting control through manipulation.

Screams of excitement shatter the stillness of Gombe after the capture of a bushbuck fawn held by the alpha male, Frodo. Once the prey is caught, the begging begins.

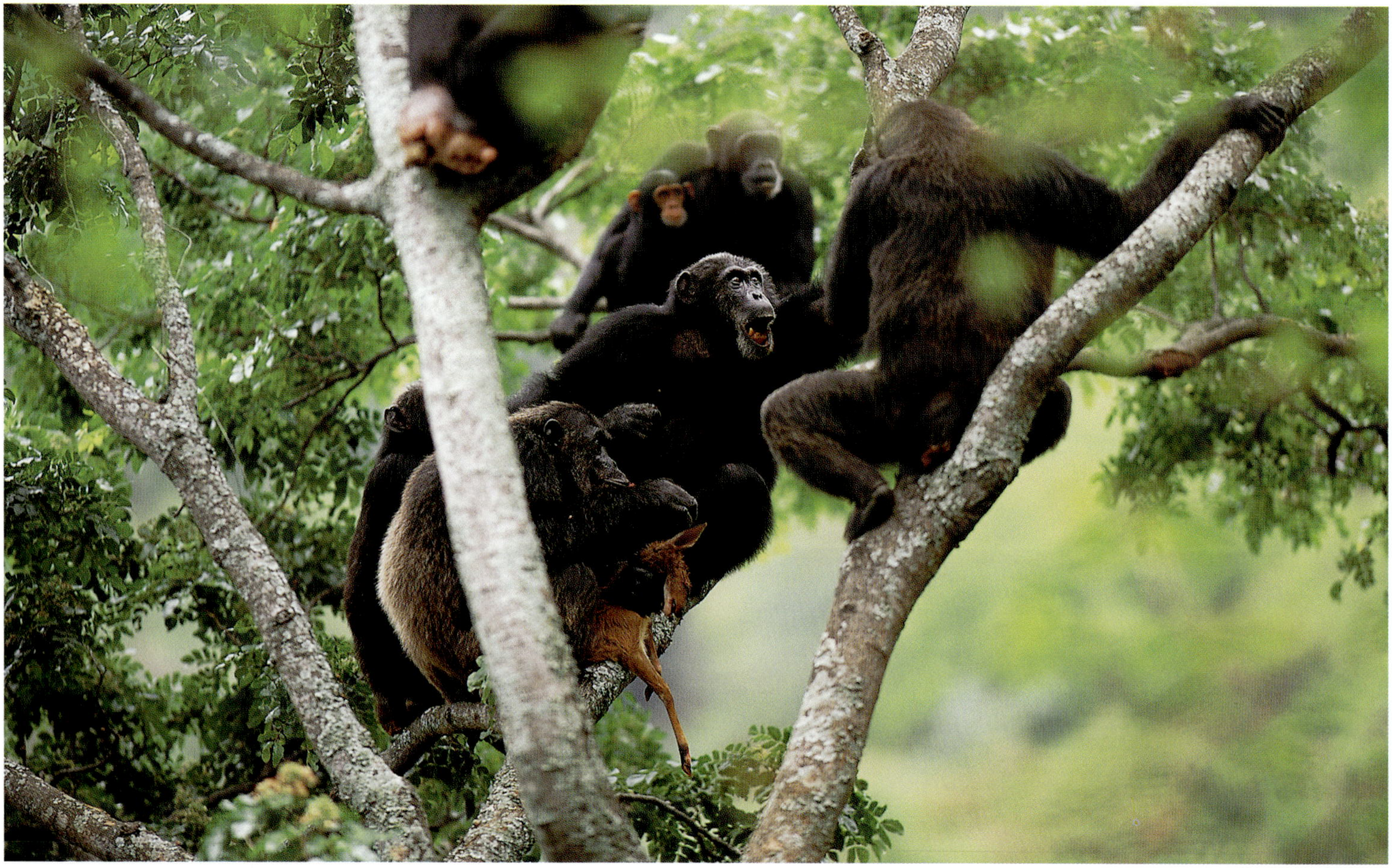

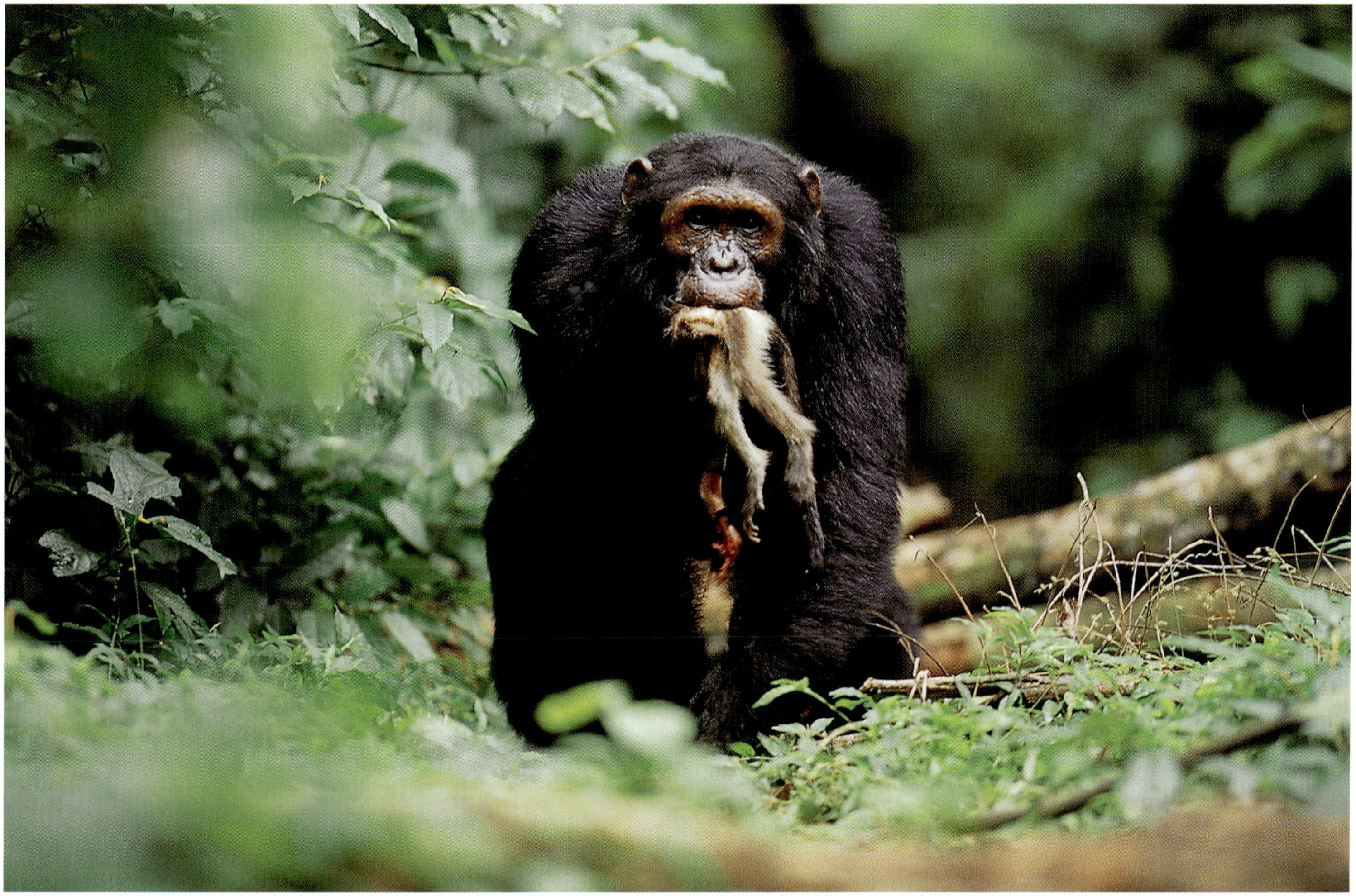

A Gombe male knuckle-walking with a captured colobus monkey in his mouth, looking for a secluded spot in which to savour his prize, away from begging chimpanzees.

Before 1974, humans were the only species known to wage war on, and even to kill, fellow beings of their own species. Then students of Goodall observed a significant event that altered this view forever: male chimpanzees silently invaded the territory of a neighbouring community, ambushed a few chimpanzees, and inflicted injury and even death. Such warfare was observed on two further occasions. The revelation of chimpanzee raiding parties suggests that early humans might have carried out such raids as well.

Toshisada Nishida's Mahale study

The moist, warm lowland forest nestled between the Mahale Mountains, 170 kilometres (105 miles) south of Gombe, and Lake Tanganyika's eastern shoreline is home to between 700 and 1 000 chimpanzees: the highest concentration of these apes on Earth. This is the site of an ongoing chimpanzee research project instituted in 1965, when the Japanese primatologist, Toshisada Nishida, arrived in Mahale to study two groups of the chimpanzees. Known as the Kyoto University Chimpanzee Project, this remains an active centre for chimpanzee research, and is still run by Nishida.

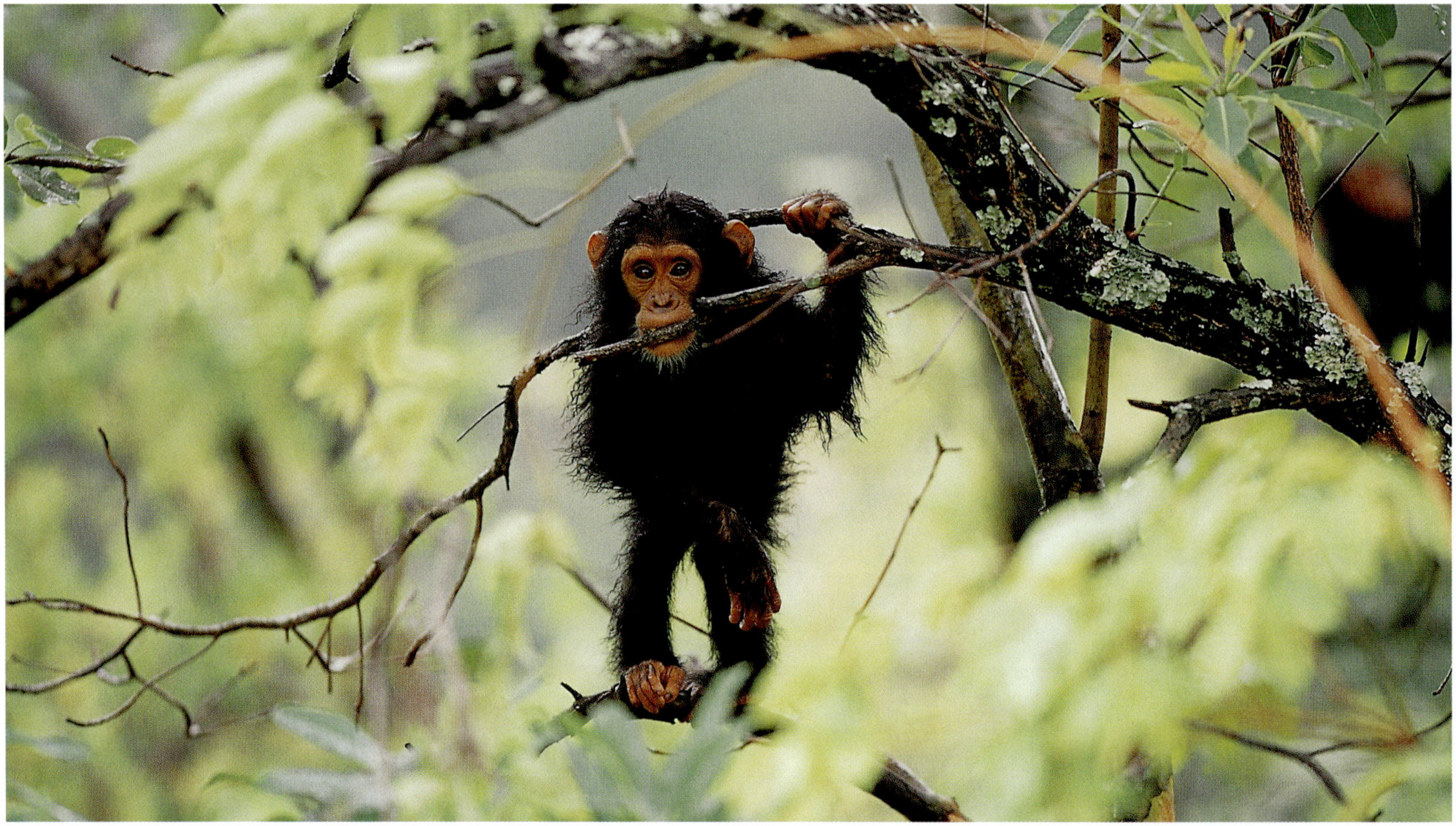

Above: *A young chimpanzee, fur wet from a recent downpour, plays alone in a tree while its mother rests out of sight on the ground below.*
Right: *Four-year-old twins, Golden and Glitta, spur their mother Gremlin on as she accelerates through a Gombe forest clearing.*

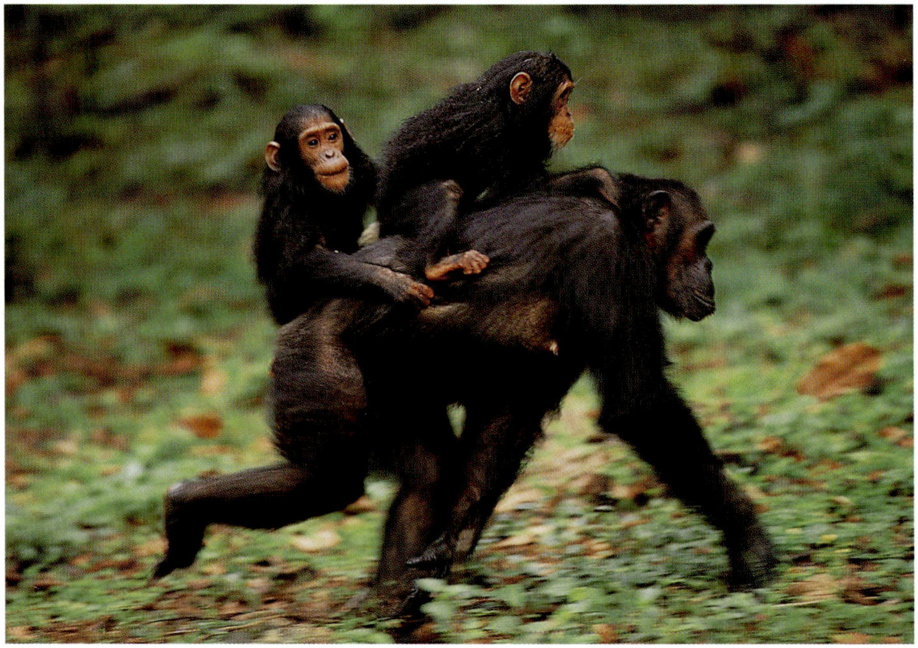

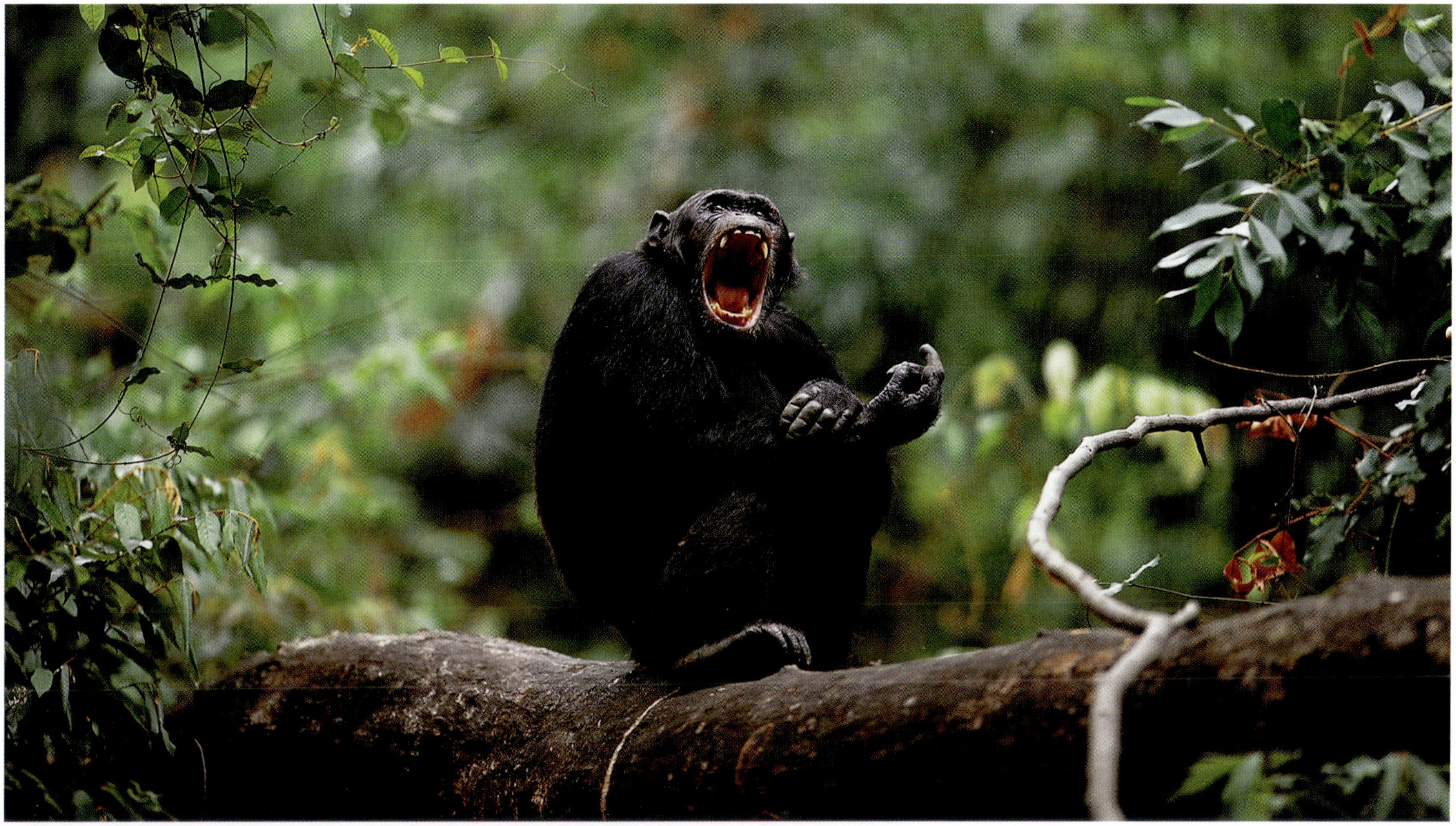

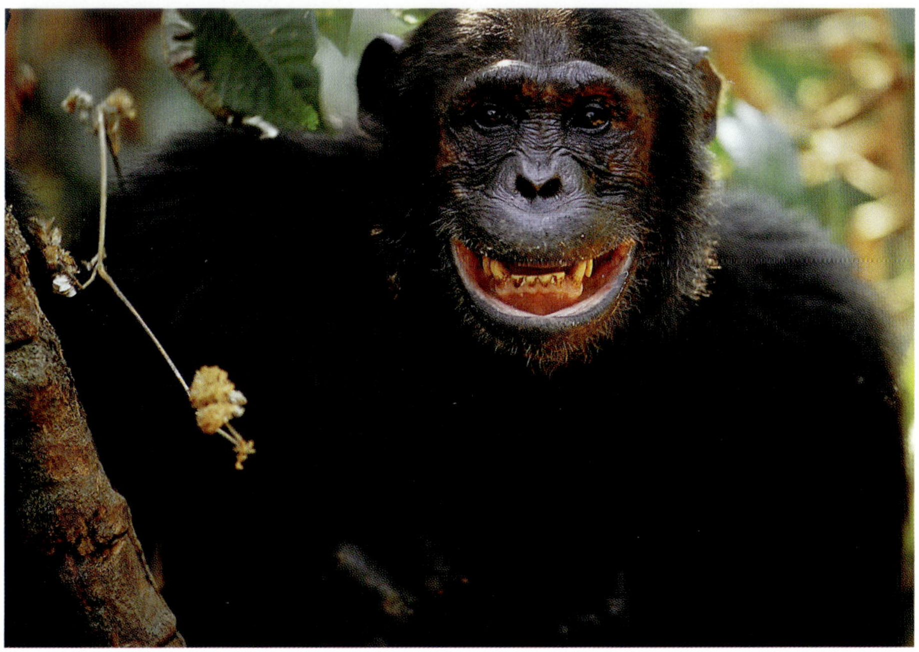

Above: *Twenty-four-year-old Gombe male, Tubbe, displays his formidable set of teeth in a wide yawn.* **Left:** *Alouf, alpha male at Mahale, breaks into an expressive grin. Chimpanzees have pliable lips and numerous facial muscles, enabling them to form a wide range of facial expressions. After humans, they have the most expressive of primate faces.*

Nishida's first breakthrough came with the realisation that chimpanzees live in communities dominated by males, rather than in social units comprising mother and child, as Goodall had contended. The males are likely to be related, and occupy positions within a male hierarchy, which is headed by the alpha male. Each community has an identity, and hostility exists between males of different communities.

Chimpanzee social structure may have had a direct bearing on the evolution of human social behaviour. When Africa's climate began to cool and the land became drier, resources became more patchy, sparse and seasonal. For chimpanzees, food resources became more dispersed (while mountain gorillas, remaining in the forests, enjoyed concentrated food resources). Along with the food, the female chimpanzees were also widely dispersed. A single male chimpanzee – unlike the forest-dwelling male mountain gorilla – was no longer able to defend all the females; territorial defence developed, and, with it, a rare phenomenon in the primate social world: male kin bonding. Within this pattern of relationships, males associate primarily with other males, and females leave their natal groups at sexual maturity. Male kin bonding also largely characterised pre-agricultural human social history.

Goblin, a former alpha male at Gombe, charges out to face a small mixed group of chimpanzees – females, their young, and low-ranking adult males – hair bristling to give an illusion of bigger size. The display is intended to intimidate his audience and assert seniority.

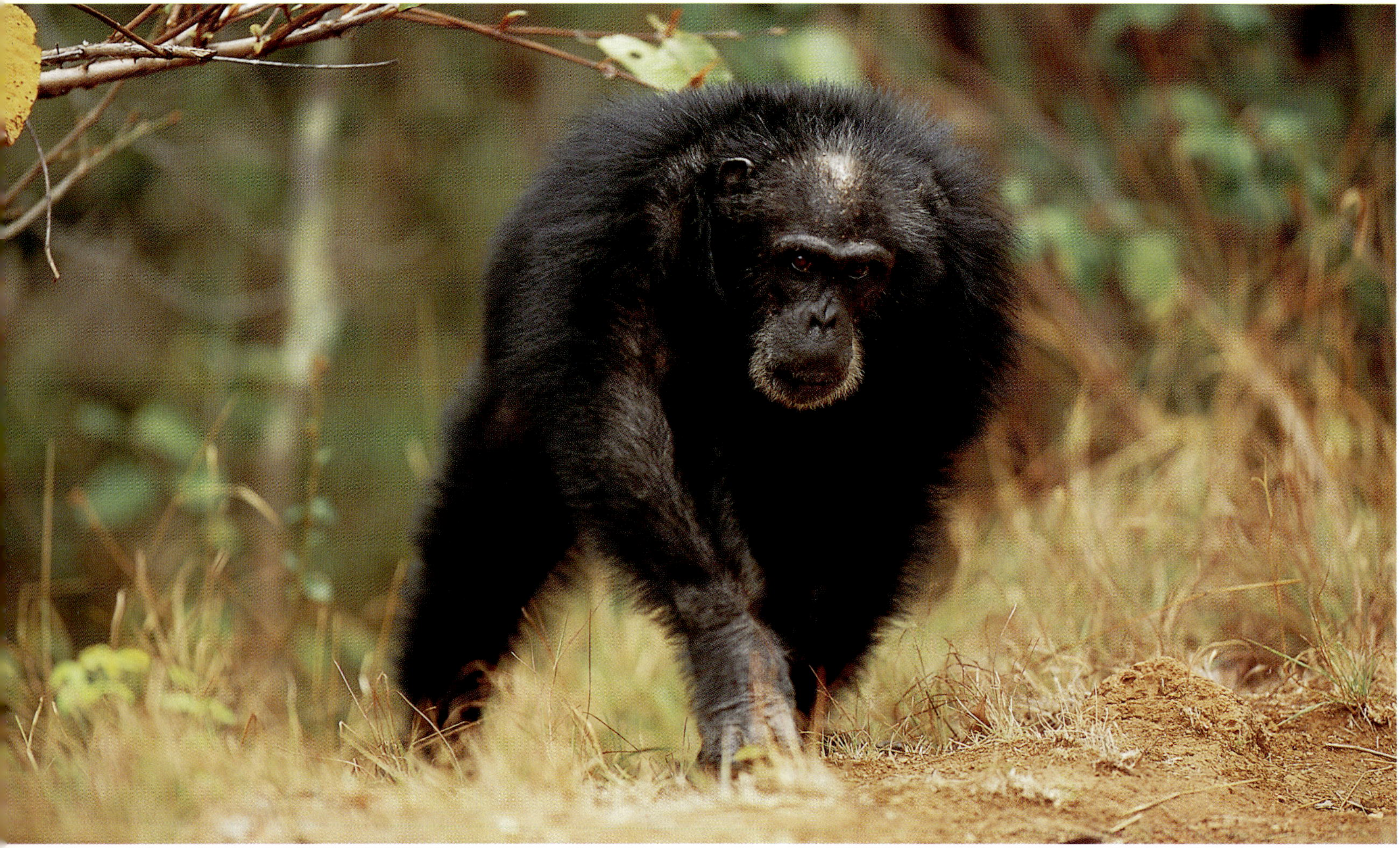

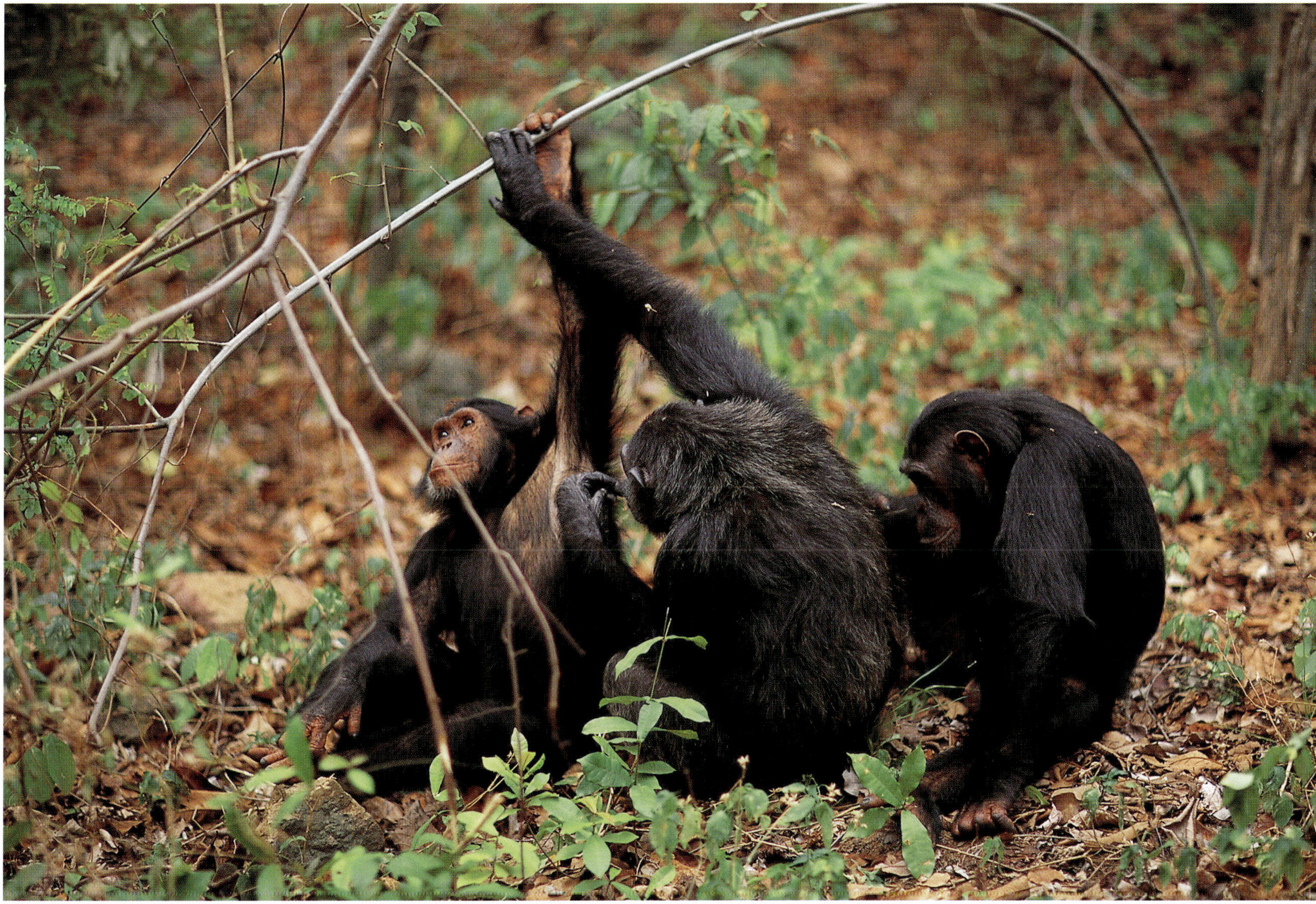

Chimpanzee culture

A topic of ongoing research at Mahale is that of chimpanzee culture. In this context, 'culture' implies the meeting of three pre-conditions: the capacity to learn and to teach; non-genetic transmission of information between and among generations; and the development of local traditions. If chimpanzees meet these pre-conditions, and can be said to have culture, then each wild chimpanzee population is defined by more than its genetic make-up. One chimpanzee population will do things differently from another, in much the same way that two villages of genetically similar tribes may have different ways of, for instance, celebrating marriage. Despite the close proximity of Gombe and Mahale, researchers have indeed uncovered some differences in chimpanzee behaviour between the two sites.

Two Gombe chimpanzees hold on to a branch while grooming one another using their free hands. This behaviour has been dubbed 'the branch clasp'. A third chimpanzee joins in the grooming session from behind.

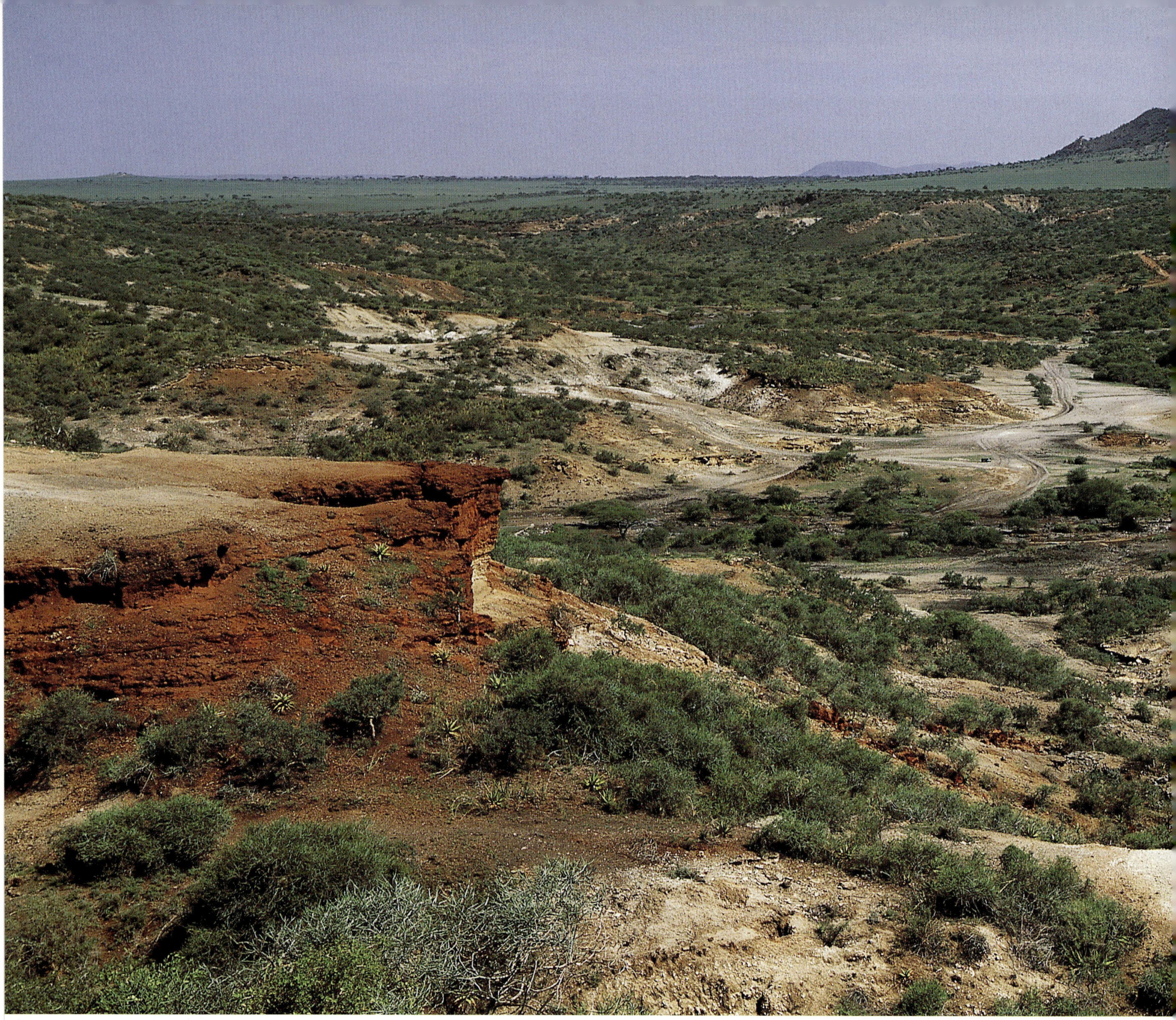

Olduvai Gorge in northern Tanzania has proved a rich source of fossils, including some early hominid remains.

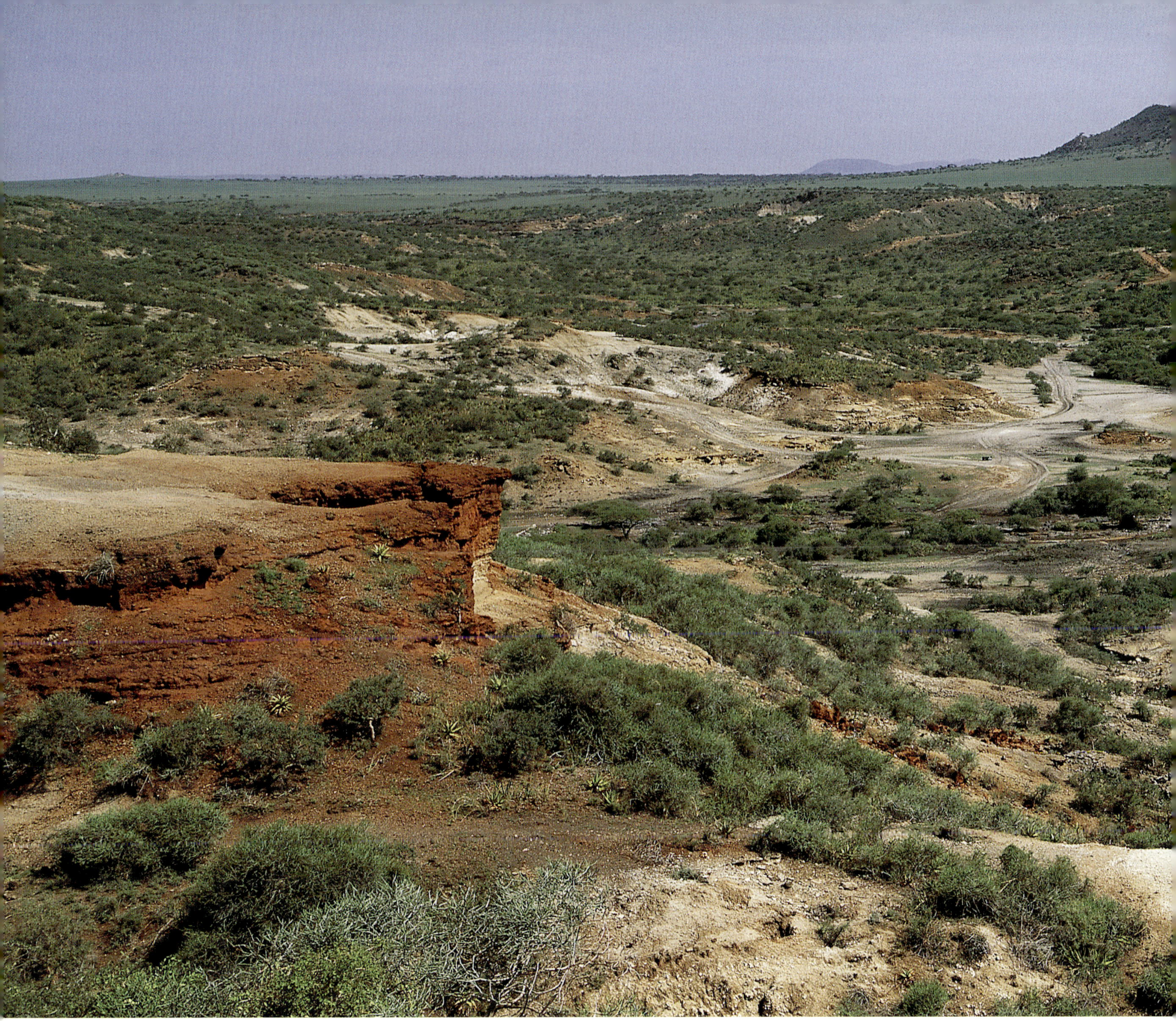

Olduvai Gorge in northern Tanzania has proved a rich source of fossils, including some early hominid remains.

Our common ancestor

While they are our closest living relatives, we did not descend from chimpanzees; instead, we are linked only by a common ancestor from which we both arose, between four and seven million years ago. Chimpanzees and humans have evolved independently for millions of years; humans, in particular, have adapted to many differing habitats and circumstances. A comparison of chimpanzee and human genetic profiles reveals that, although the two genomes differ by only 1.2 per cent, that difference accounts for 35 million nucleotide modifications (parts of the DNA structure that have changed) since humans and chimpanzees diverged from the common ancestor, millions of years ago. It is pointless to compare modern humans with chimpanzees, although it may be useful to look at today's chimpanzees in order to understand the lives of the prehumans of six million years ago; chimpanzees may not have undergone as many changes as humans have, since their habitat has not changed much.

If modern chimpanzees are to serve as the model, our early ancestors lived in societies with many related males, all competing for the females in their group. They had a taste for meat and developed rudimentary skills for fashioning very simple tools. There was some variation from population to population in how they did things; neighbouring populations sometimes waged war. Then, as the climate became drier, some of our ancestors left the trees to explore their new environment; they developed upright walking, hunted, migrated, discovered fire; the differences between them and the chimpanzees grew.

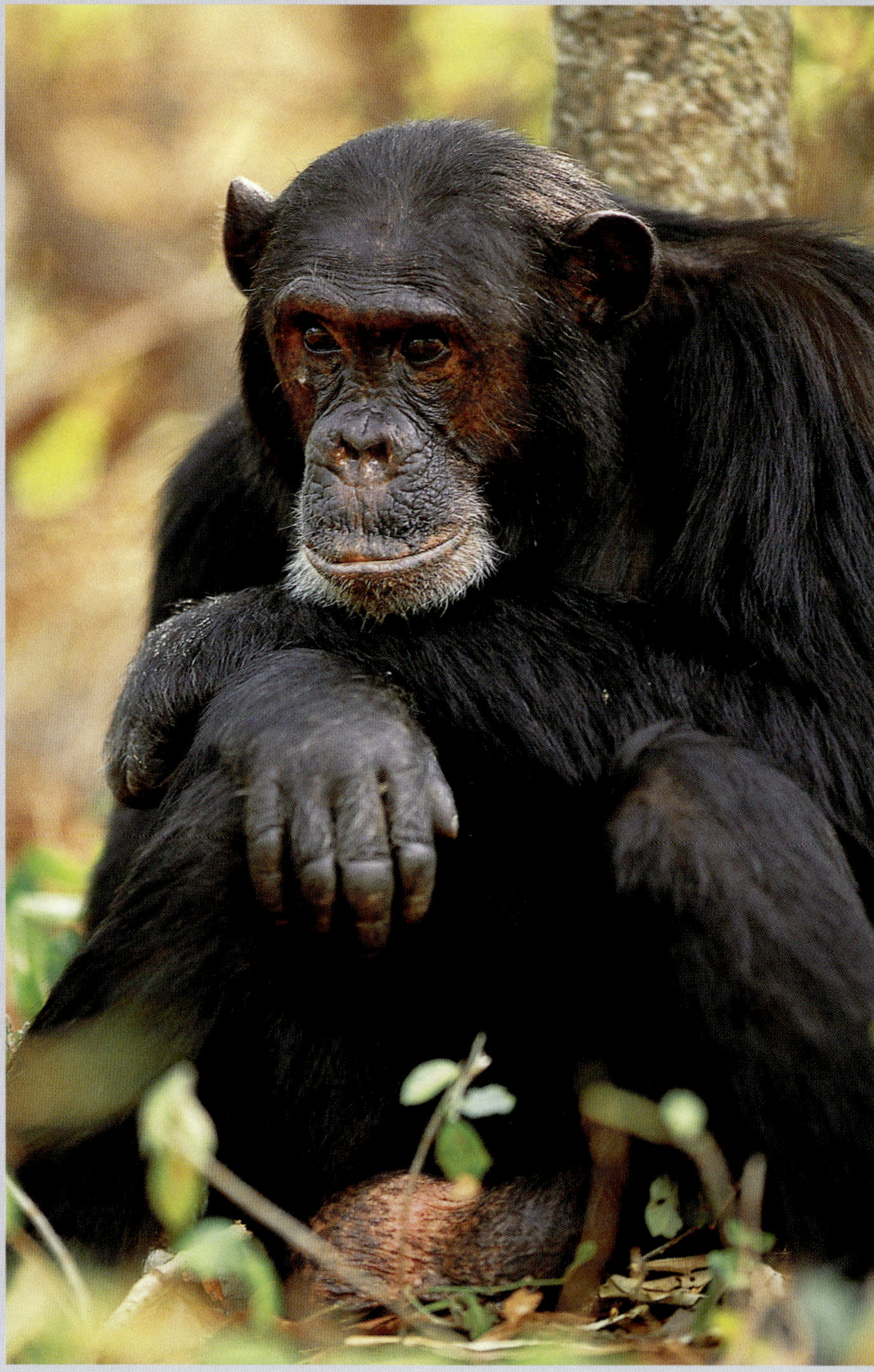

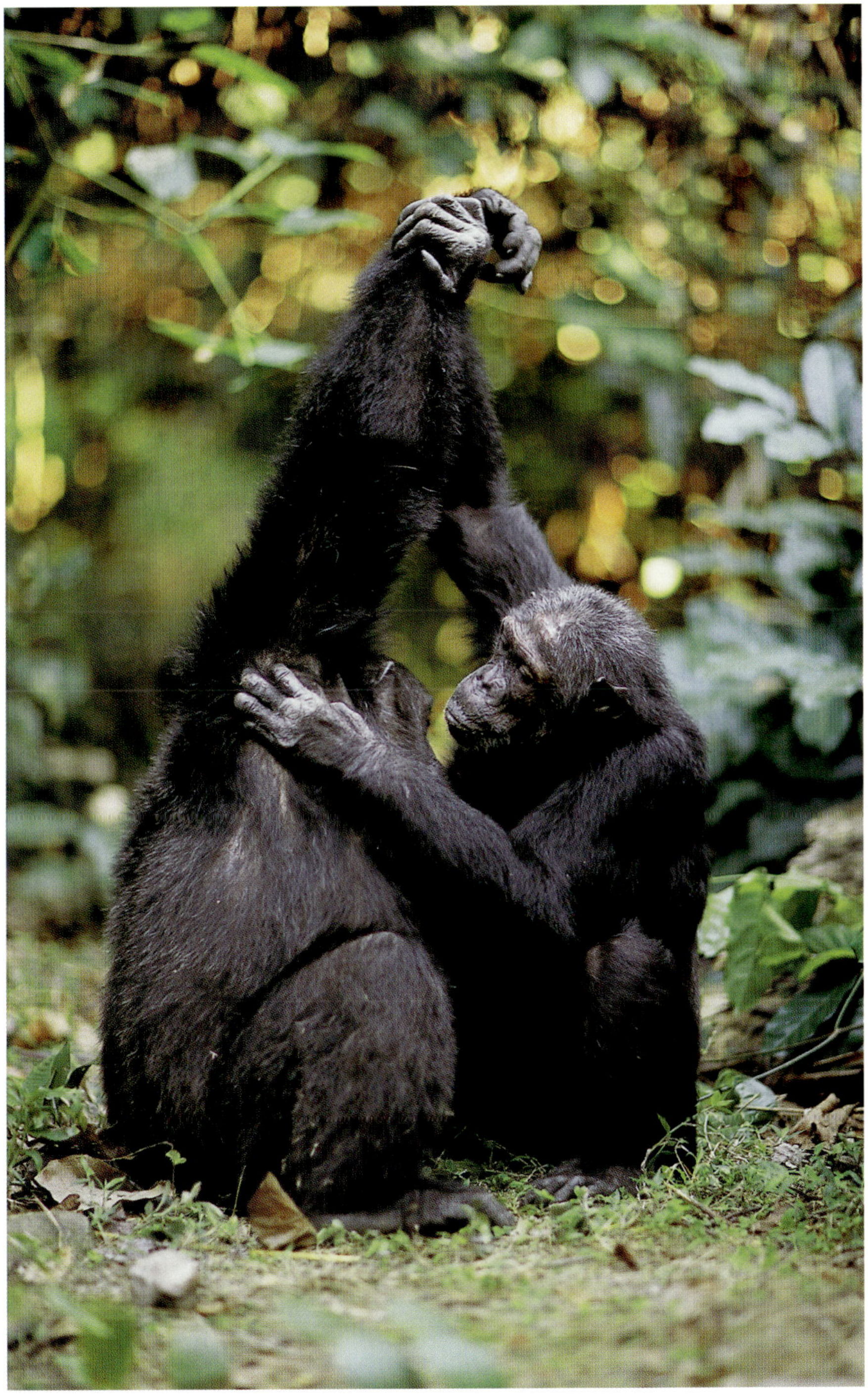

Performing double-handclasp grooming, the chimpanzees simultaneously extend an arm overhead and then grasp the other's wrist or hand. The opposite hand is used to groom the other individual's underarm area revealed by the raised limb.

The food choices of the two groups demonstrate differences: of 286 potential food types common to both Gombe and Mahale, only 104 are eaten by both groups; at Gombe, the fruit of the oil-nut palm, for instance, is the single most important food item, while the Mahale chimpanzees ignore it completely. There are also variations in the way food is collected and prepared, and in the tools used. Gombe chimpanzees bang hard-shelled fruit against tree trunks or rocks, while Mahale chimpanzees use their teeth; drinking with the help of leaf sponges is common practice at Gombe, but has been seen only once at Mahale; termites and tree ants are found in both Gombe and Mahale but, unlike Gombe chimpanzees, those of Mahale never use sticks to fish for termites, using their fingers instead to crumble apart the soil and pick out the delicacy.

Courtship signalling by males is performed differently by the two groups. At Gombe, courtship involves a male's rapidly shaking a small bush or branch several times, after which a female in the area may approach the male and demonstrate her sexual receptivity by presenting her genital swelling to him. In Mahale, males will pluck a leafy stem from a nearby plant and noisily use their teeth and fingers to tear off its leaves. Directed at a particular female, this signals a male's desire.

Differences between the two populations are also apparent in the way they perform grooming. At Mahale, two grooming partners perform double-handclasp grooming: by holding each other's hands, their arms held straight up, each uses the free hand to groom the grooming partner's exposed armpit and side. At Gombe, grooming chimpanzees use a different technique, each grasping an overhead branch for support.

The chimpanzees at the two sites, only 170 kilometres (105 miles) apart, clearly display some cultural diversity; the differences in behaviour between the two populations have no genetic basis, and are due entirely to differing cultural traditions.

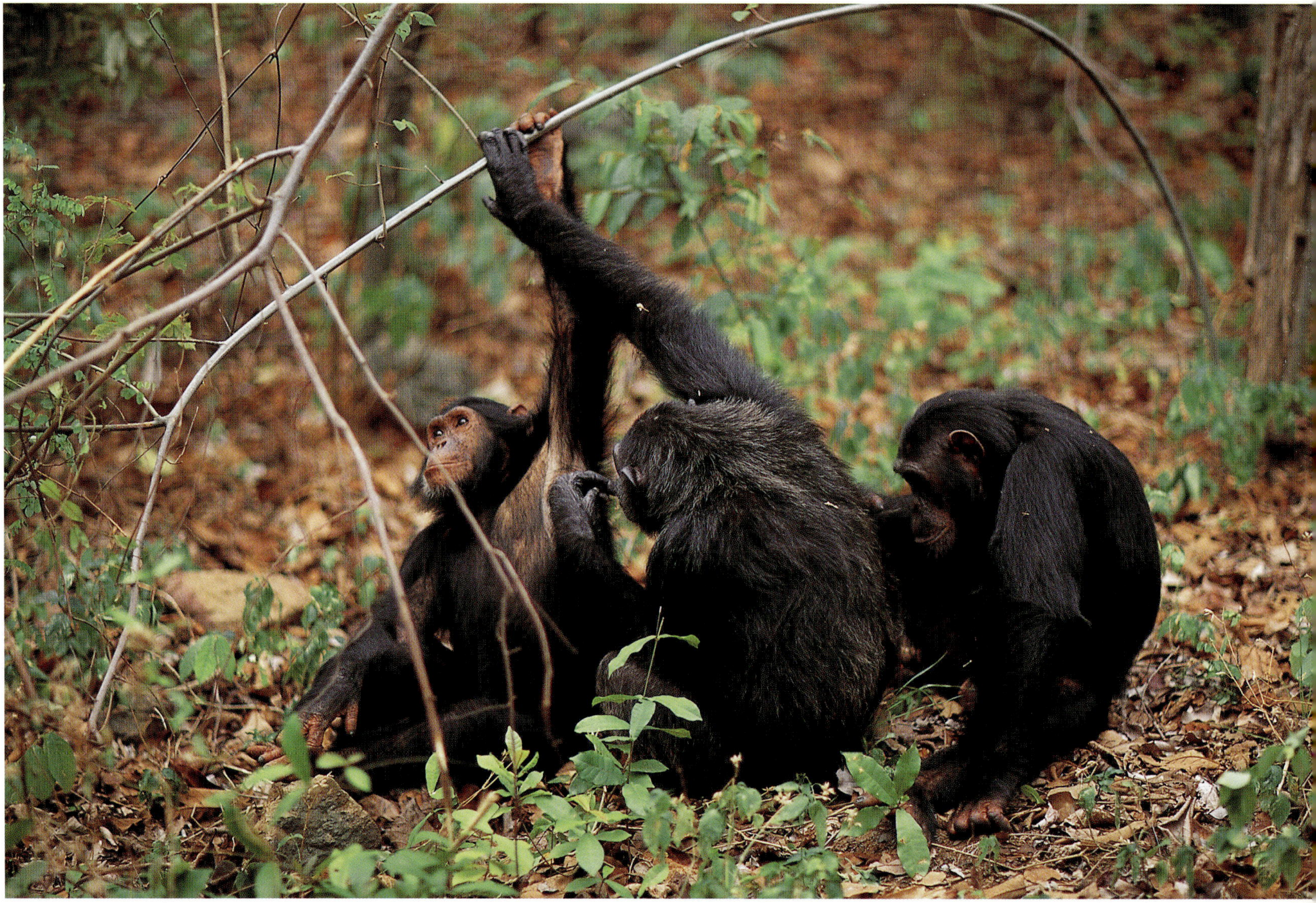

Chimpanzee culture

A topic of ongoing research at Mahale is that of chimpanzee culture. In this context, 'culture' implies the meeting of three pre-conditions: the capacity to learn and to teach; non-genetic transmission of information between and among generations; and the development of local traditions. If chimpanzees meet these pre-conditions, and can be said to have culture, then each wild chimpanzee population is defined by more than its genetic make-up. One chimpanzee population will do things differently from another, in much the same way that two villages of genetically similar tribes may have different ways of, for instance, celebrating marriage. Despite the close proximity of Gombe and Mahale, researchers have indeed uncovered some differences in chimpanzee behaviour between the two sites.

Two Gombe chimpanzees hold on to a branch while grooming one another using their free hands. This behaviour has been dubbed 'the branch clasp'. A third chimpanzee joins in the grooming session from behind.

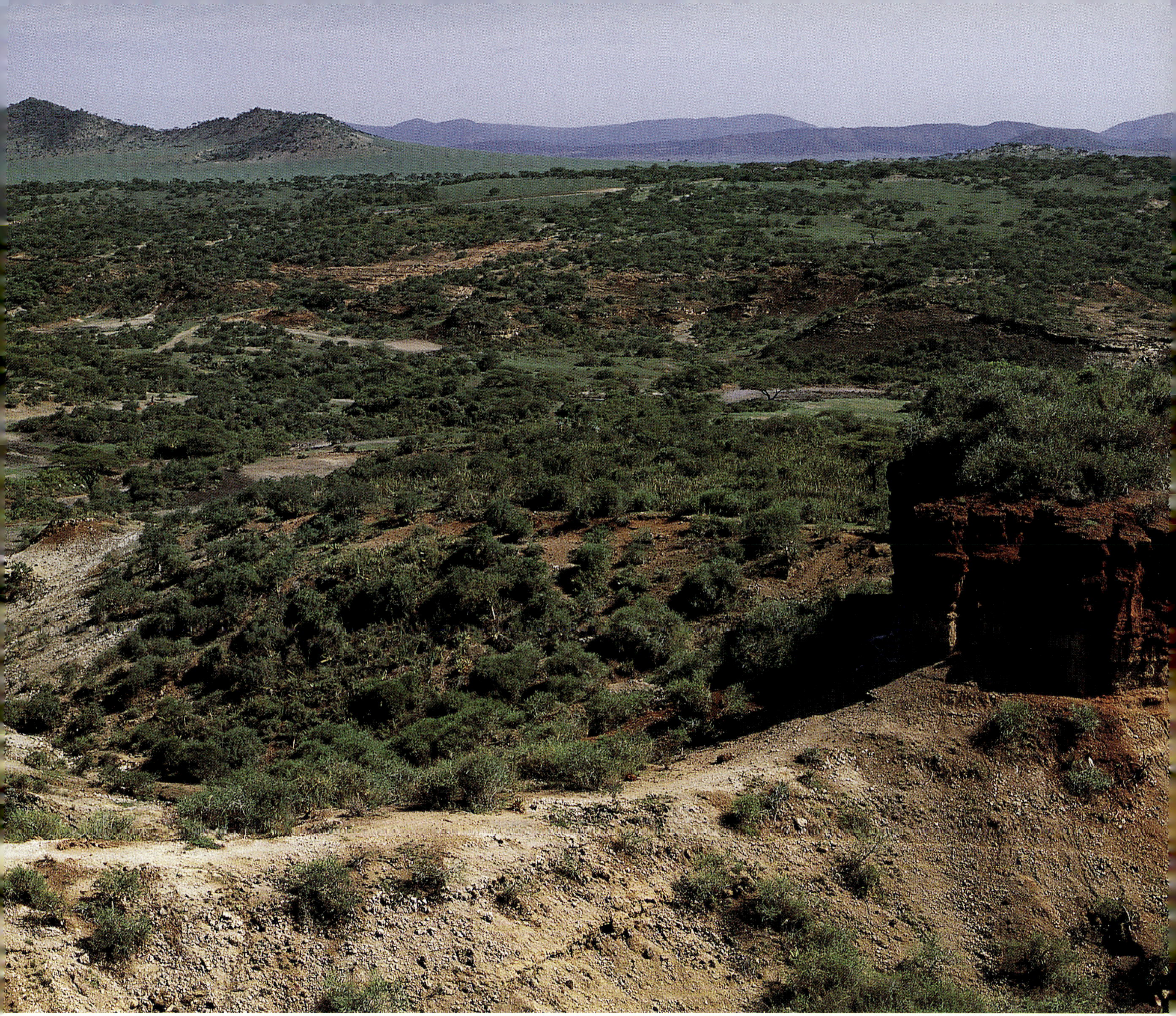

7 Prehumans of the ancient lakes

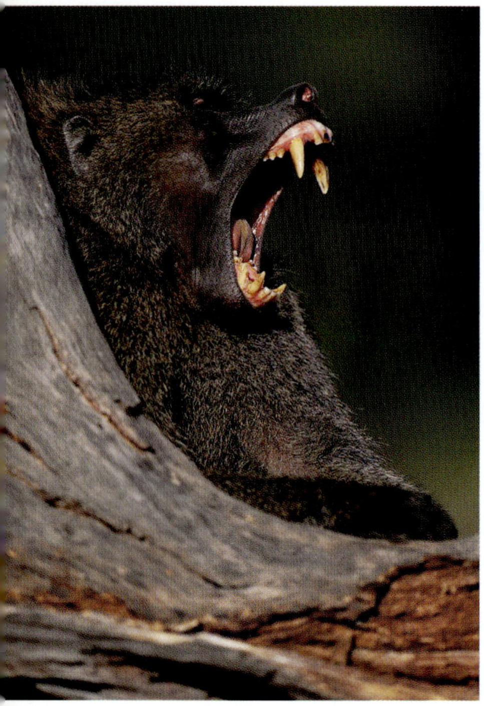

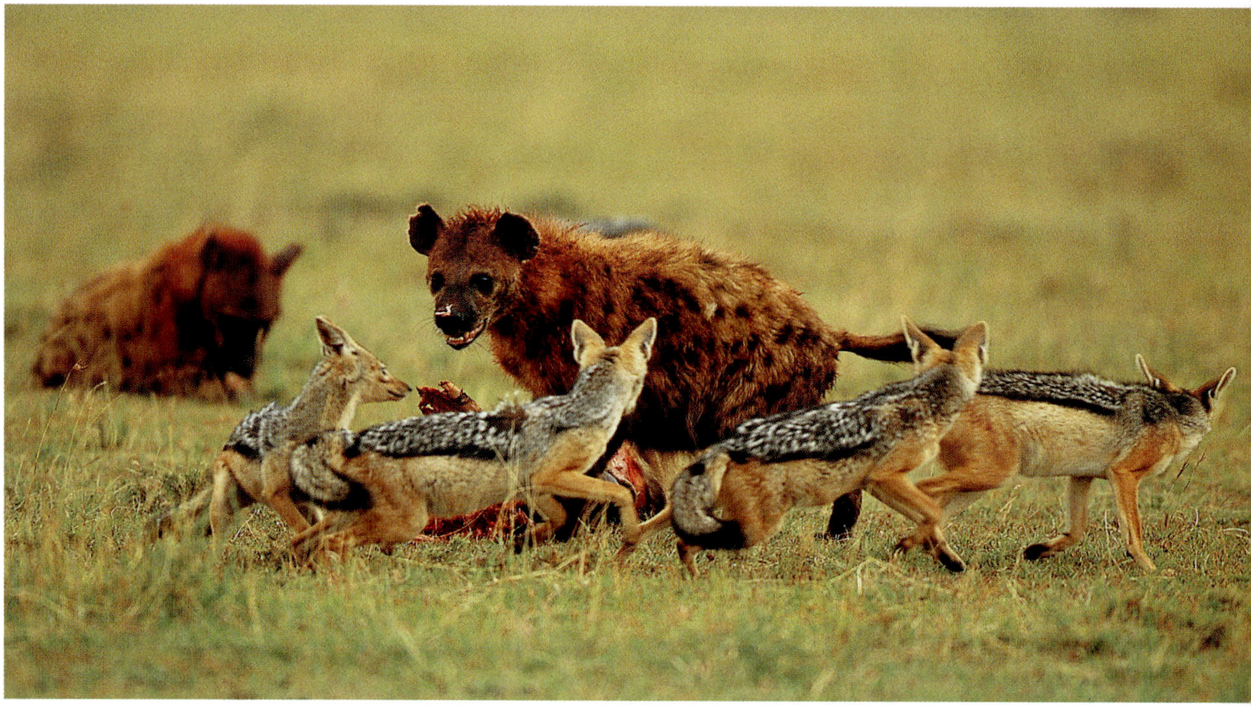

Above: *Baboons and hominids are two groups of primates that have extensively adapted to open terrestrial environments in East Africa. Hominids, lacking the baboons' powerful teeth, compensated by developing larger brains.*
Above right: *It is laughable to think of prehumans – not fast runners and lacking fearsome sets of teeth – as fierce hunters. Like jackals,* Canis melampus, *here circling around the kill of a hyaena,* Crocuta crocuta, *they most probably supplemented their vegetarian diet by scavenging for meat.*

'In each great region of the world, the living animals are closely related to the extinct species of the same region. It is therefore probable that Africa was formerly inhabited by extinct apes closely allied to the gorilla and the chimpanzee; and as these two species are now man's nearest allies, it is somewhat more probable that our earliest progenitors lived on the African continent than elsewhere.'

So wrote Darwin in 1871 in his book, *The Descent of Man*. And, as Darwin's ideas spread, the hunt for a single 'missing link' between humans and apes began. At that time only one prehuman fossil species was known – *Homo neanderthalensis*, or 'Neanderthal Man' – although since then, over almost 140 years of dedicated searching, many more prehuman fossils have come to light.

One person who was determined to find the missing fossil link was Louis Leakey (1903–72). Born in Kenya and educated at Cambridge, Leakey was determined to make his name by fulfilling Darwin's prediction.

The rift valley as a fossil bank

Leakey was convinced that the rift valley's geological profile was critical to the discovery of prehuman fossils, and later exploration vindicated this belief. Tectonic movements and the resulting volcanoes in the rift valley created a string of closed lake basins that continually shifted under the dynamic conditions of the rift, leading to the build-up and subsequent exposure of large banks of lake sediments. The gentle deposition of sediments in the lakes was conducive to the preservation of animals that died on their shores, and faulting and erosion later exposed these fossils, several million years later, for palaeontologists to discover. An important factor in the success of the process was the presence of large quantities of volcanic deposits rich in potassium, a medium suitable for accurate dating using the potassium-argon dating method.

Olorgesailie

Lying along the baking rift valley floor, 65 kilometres (40 miles) from Nairobi, is an ancient site called Olorgesailie, which once stood beside a large lake. In 1918, long after the lake had disappeared, geologist John Gregory was scouting the area for mineral prospects when he came across a stretch of open grassland littered with anomalous dark stones that had clearly been shaped by human hands. Despite Gregory's discovery, Olorgesailie was left undisturbed for more than two decades afterwards; only in 1942 did Louise Leakey eventually reach the site. He and his wife Mary began the excavations that are ongoing today. The Leakeys uncovered a site where tools had been manufactured in huge numbers for roughly one million years, between 1.2 million and 200 000 years ago. There were hundreds of large, well-preserved Acheulean-style hand-axes, cleavers and round stones (associated with prehuman industry), as well as broken animal bones. The bones and skulls included the remains of a number of extinct animals.

Quartz and obsidian rocks, such as those from which the axes were made, have never been found on the valley floor. They would have been carried in from the surrounding hills, as far as 10 kilometres (6 miles) away – a long way to carry an armful of stones. The excavations revealed that in some locations, axes were fashioned, while at others, blunt implements were brought in for re-sharpening. Olorgesailie was a factory with a repair department, one that stayed in business for a million years.

Amid all these tools, stones and animal bones Leakey was, to his great disappointment, never able to find any prehuman or human fossils. It seems that, however much time Olorgesailie people spent at the site, they went elsewhere to die. Then, about 200 000 years ago, the lake dried up and the people moved away. Despite over 60 years of searching, we have very little idea of who they were, or where they went.

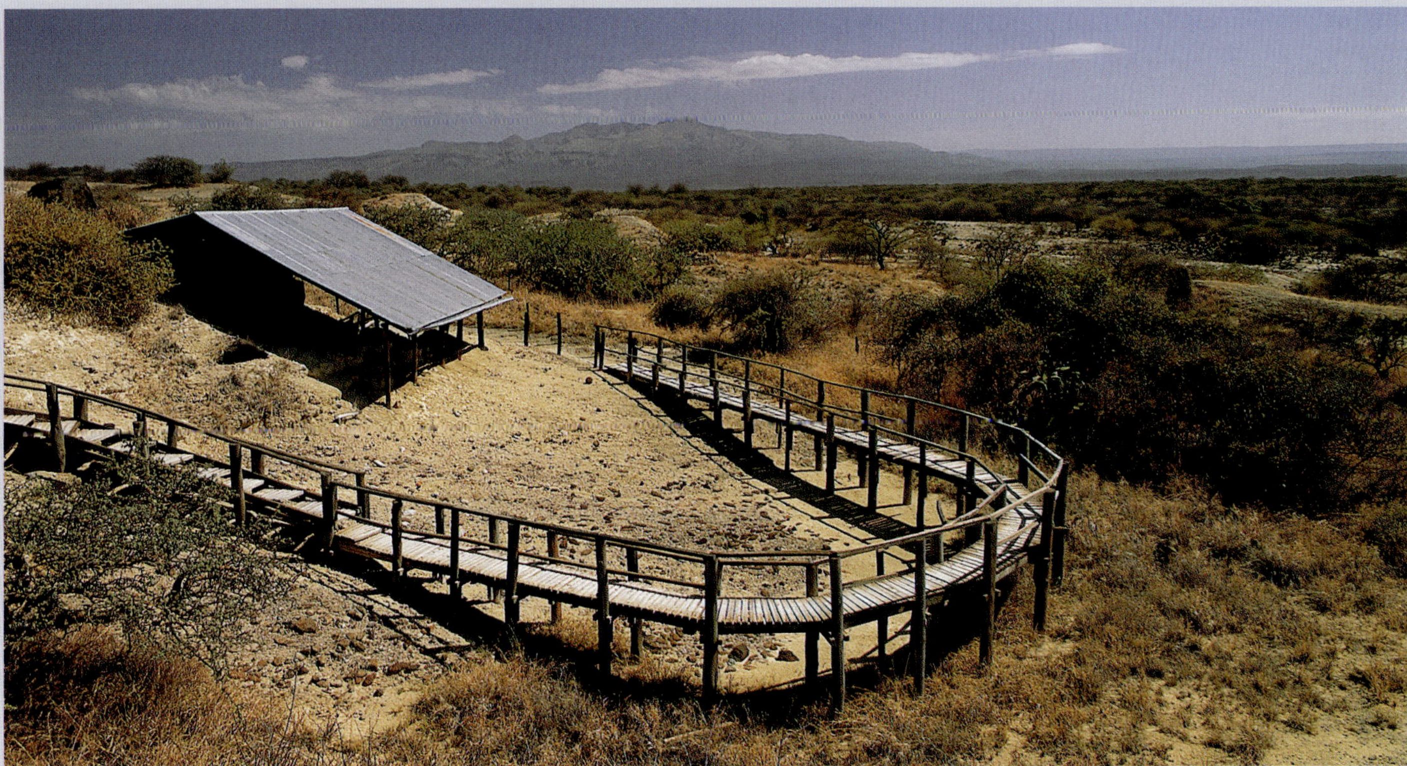

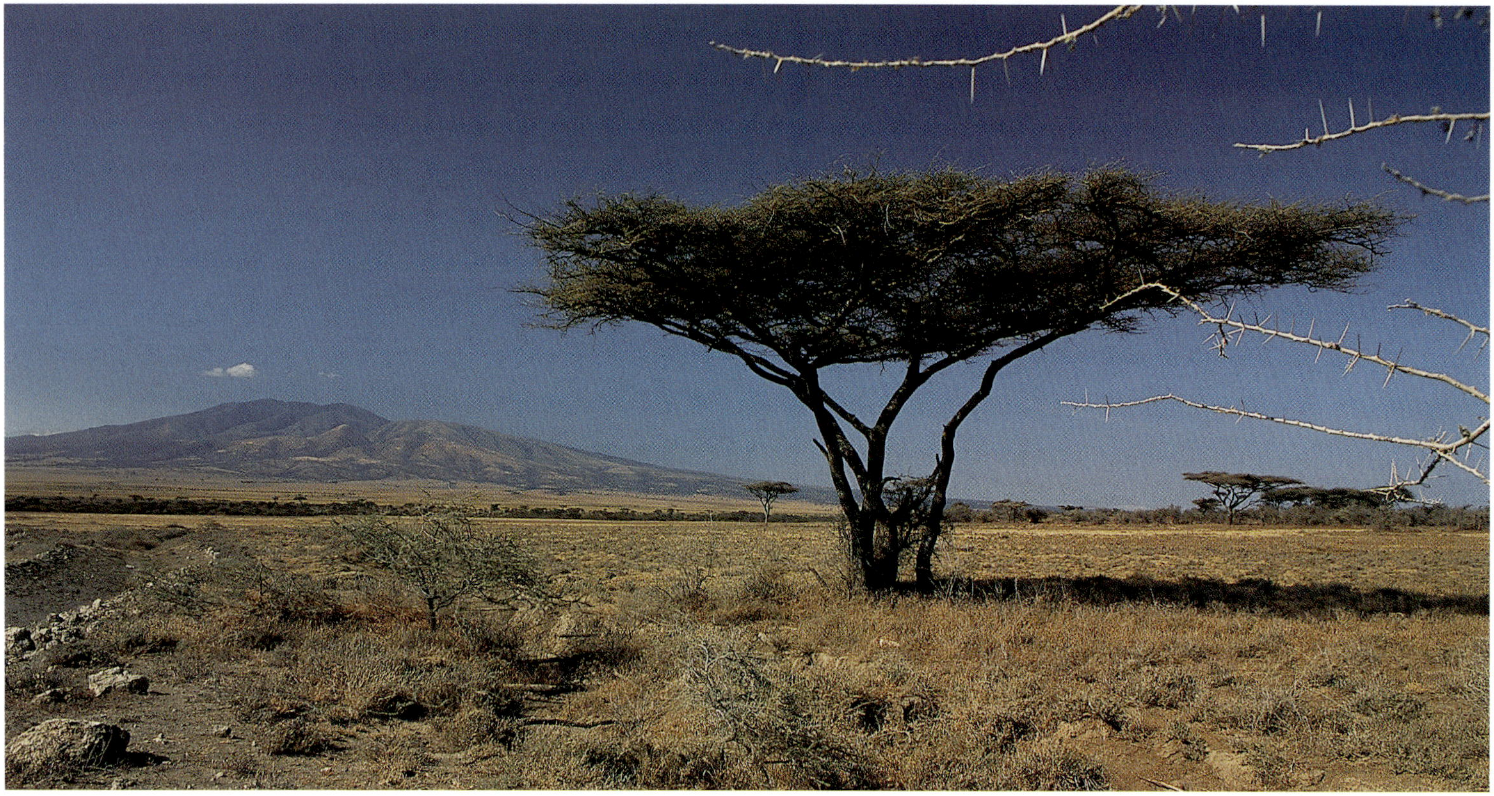

Open savannah habitats, such as that around Olduvai Gorge, are thought to have provided the context in which much of early hominid evolution occurred.

Olduvai Gorge

Over the aeons, the widening rift provided a receptacle for rich deposits of fossils, the most extensive and diverse of which come from Olduvai Gorge in Tanzania. From the top to the bottom of this ravine, which is 100 metres (328 feet) deep, a record of life goes back, layer by layer, nearly two million years.

The Leakey family's association with Olduvai Gorge began in 1931 when Louis accompanied Hans Reck to the gorge. At the site, Reck had discovered fossil bones and a human skeleton, and identified four layers of volcanic deposits, labelled Beds 1 to 4, ranging in age from two million years to about 17 000 years.

For nearly 25 years, a lack of funds prevented Leakey and his family associates from carrying out sustained research into these rich deposits, although they visited the gorge whenever they could. Their work convinced them that many prehumans had lived on these plains, beside a lake that had covered much of the region at the time, but which had long since vanished. Yet for years the key to the identity of the prehumans living at Olduvai eluded them.

The breakthrough came in 1959 when a fragment of bone protruding from the wall of Bed 1, the oldest layer, proved to be part of the skull of a prehuman. Over the next 18 months, the Leakeys and their African assistants dug out and pieced together some 400 fragments of what came to be called 'Nutcracker Man', *Paranthropus boisei*. Modern dating techniques suggest that the individual lived 1.85 million years ago. Much to Leakey's disappointment, Nutcracker Man was not a direct ancestor of *Homo sapiens*, and thus did not belong to the *Homo* line leading to humans, but rather was a member of a side branch, called 'southern apes', which disappeared 1.5 million years ago.

Nevertheless, Nutcracker Man brought fame to Olduvai Gorge. The next major find, made by one of Leakey's sons, Jonathan, was assigned to the species *Homo habilis*, or 'Handyman', the species probably responsible for making some of the tools found at Olduvai. Handyman was more ape than human, but his brain size was considerably larger than that of Nutcracker Man. Conventionally, the *Homo* line begins with Handyman, *Homo habilis*, and concludes with humans, *Homo sapiens*; but in between the two, there have been half a dozen or so other *Homo* species.

Lake Turkana

It had not occurred to scientists to excavate the sedimentary deposits bordering Lake Turkana, but a chance occurrence revealed its potential as a productive site for early human remains. In 1969 Richard Leakey, another of Louis Leakey's sons, and Maeve Epps came upon a prehuman skull lying on the surface of a dried-up river bed. The skull belonged to *Homo habilis*, the same species as Handyman of Olduvai Gorge, excavated 10 years before.

Lake Turkana was to yield up further treasures: another *Homo habilis* skull was found. Labelled '1470', it was older than the Olduvai specimen, dating back more than two million years. Then, in 1984, Richard Leakey's team found a nearly complete *Homo erectus* skeleton. They called it 'Turkana Boy', and estimated that it had lived 1.54 million years ago, and died aged between nine and 12 years. It had an entirely modern body structure. Although *Homo erectus* had been known about for almost a century – a Dutch scientist had discovered the first specimen in Java in 1891 – it was known only from scattered fragments; this specimen was an almost complete skeleton.

The site of Japanese palaentological excavations near Lake Turkana in the rift valley, northern Kenya. The search for hominid remains has, to date, delivered limited but encouraging results.

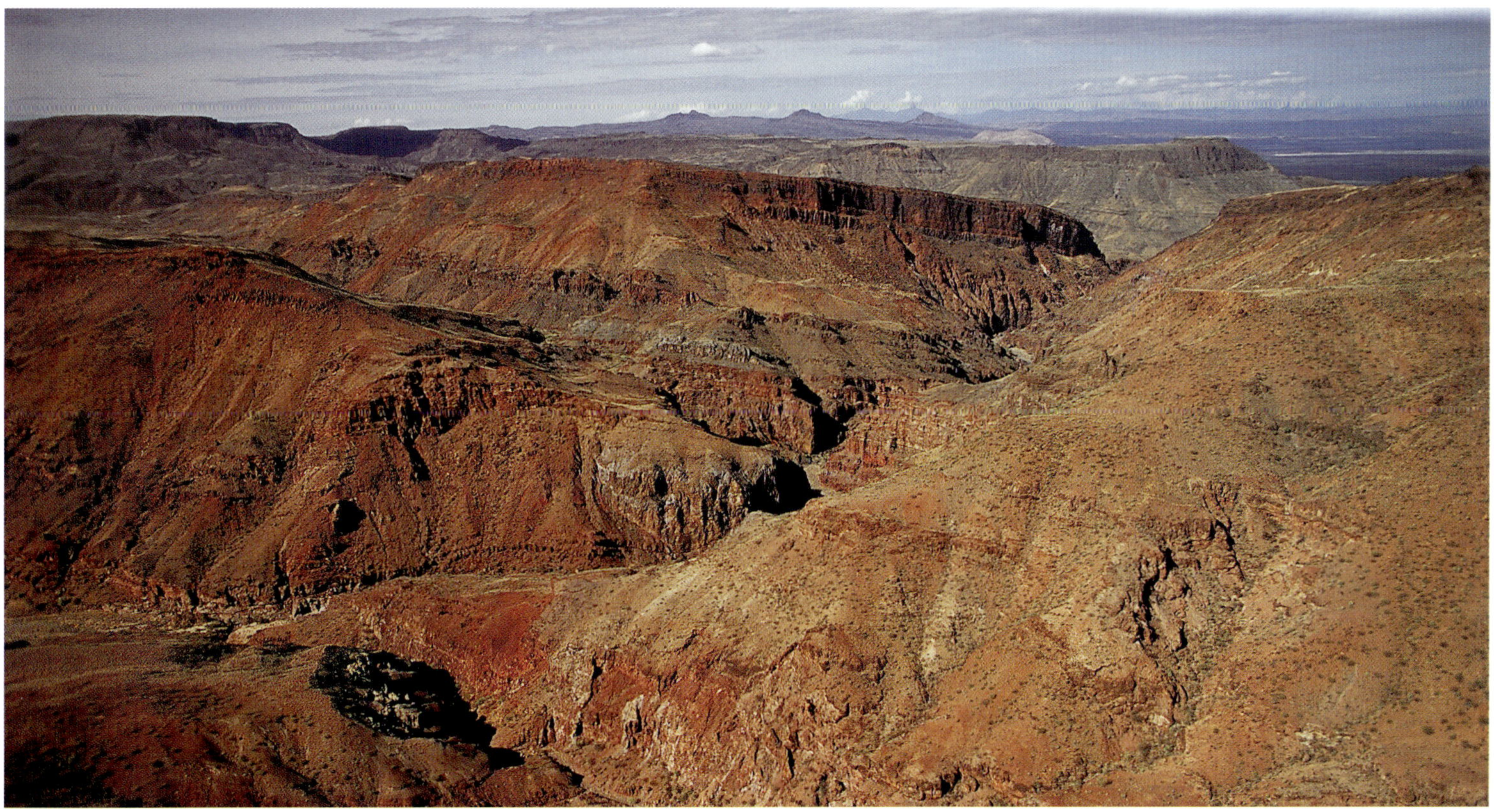

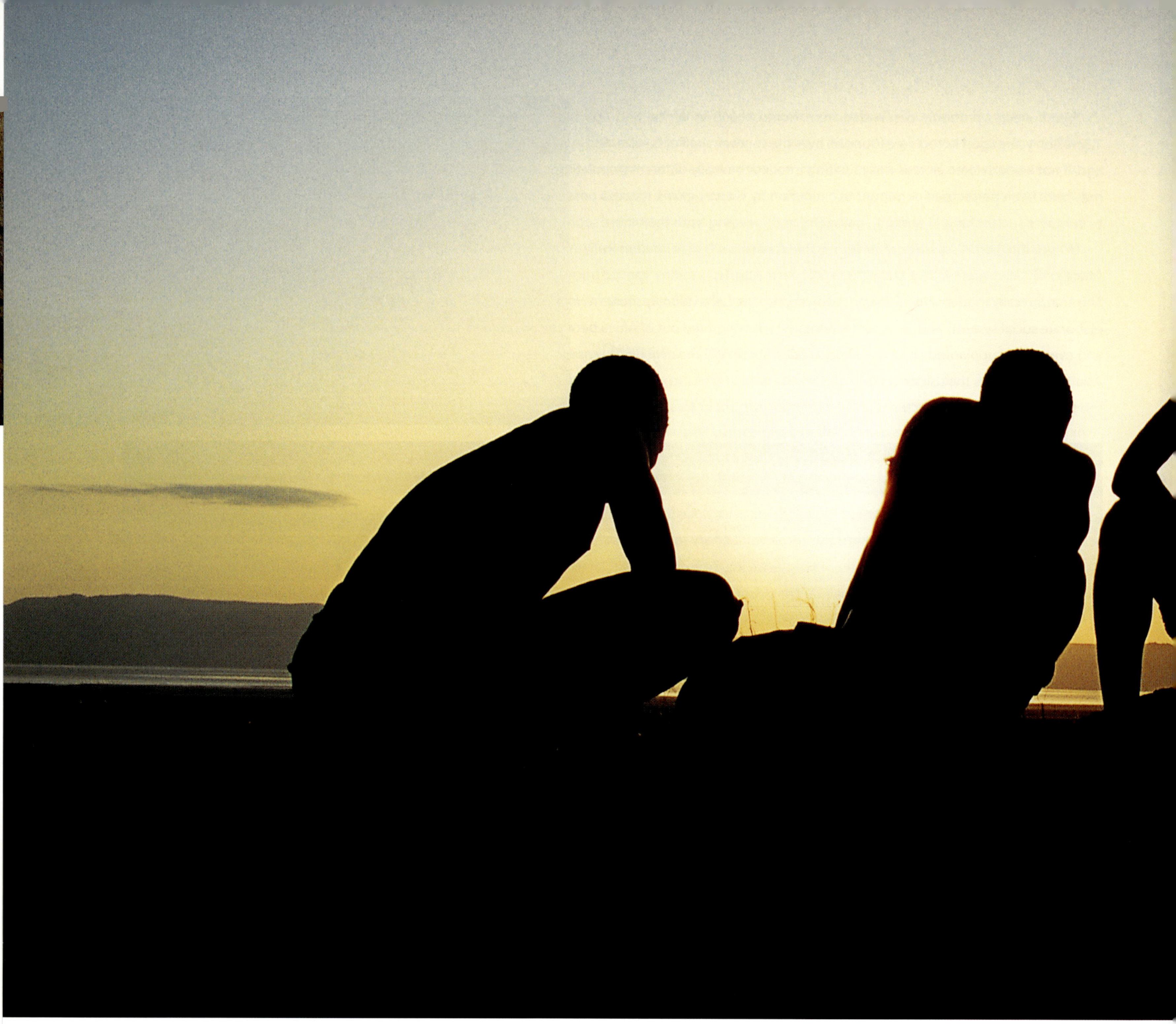

Hadza hunters scan for game from a hill-top vantage point. They use handmade bows and arrows, and are highly skilled at tracking game.

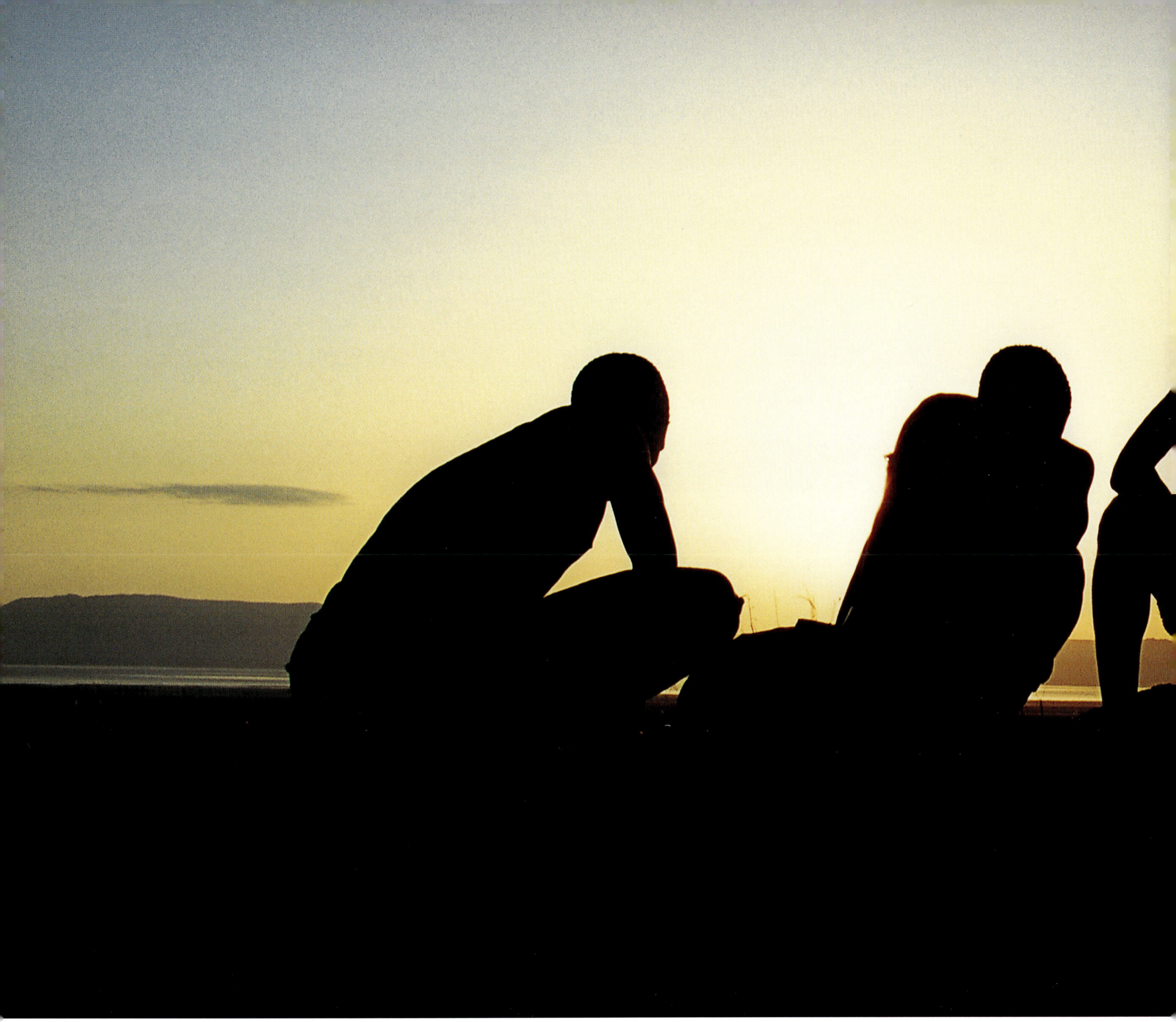

Hadza hunters scan for game from a hill-top vantage point. They use handmade bows and arrows, and are highly skilled at tracking game.

Homo erectus probably descended from *Homo habilis* in Africa, and flourished in the rift valley about 2 to 1.5 million years ago. His tools are found all over the southern half of Eurasia, and in Africa. It is clear that 500 000 years ago, if not earlier, *Homo erectus* was radiating into anatomically different populations, some of which died out. They may have been persecuted or hunted to extinction by *Homo sapiens*; it is also possible that they were simply unable to compete successfully, in hunting, gathering and breeding, with their more sophisticated rivals, *Homo sapiens*.

Where did *Homo sapiens* come from? The consensus is that anatomically modern *Homo sapiens* arose from Africa's rift valley around 250 000 to 200 000 years ago. Its ancestry can be traced back to the rift's *Homo erectus*. The anatomically modern *Homo sapiens* evolved into the behaviourally modern *Homo sapiens* (with complex culture, elaborate social systems and superior technology) who migrated out of Africa between 80 000 and 50 000 years ago, and eventually supplanted all other *Homo* species, such as *Homo neanderthalensis*, or Neanderthal Man. (The 'out of Asia' version argues that, since we don't know the direction of the migratory routes, the evidence is consistent with the possibility that *Homo erectus* evolved into *Homo sapiens* in Asia, and then migrated to Africa and Europe.)

Evidence from molecular biology

Over the past 25 years a revolution has taken place in molecular biology that has produced new insights into evolutionary relationships. Comparisons of the building blocks of life – proteins, amino acids and the sequences of DNA that make up the genetic code – have destroyed some long-held beliefs. Principal among these is that all apes – the gibbons and orangutans of Asia, the chimpanzees and gorillas of Africa – are closely related to each other. Despite similarities in anatomy and behaviour, apes do, in fact, separate into Asian and African groups – and humans belong with the African branch. More surprising is that chimpanzees are more closely related to humans than they are to gorillas. The genetic patterns found by the ongoing Global Genographic Project, which gathers DNA from populations around the world to retrace patterns of prehistoric human migration, favour the 'out of Africa' interpretation of prehistory, which contends that modern humans originated in the rift valley.

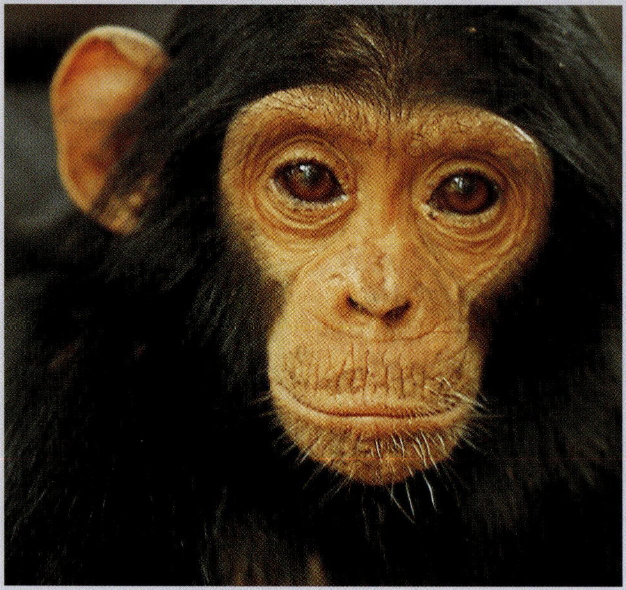

Darwin's contemporaries looked for the 'missing link'. Today, as prehuman fossils continue to be discovered, revealing further complexity in our origins, palaeoanthropologists talk in terms of several 'missing links', and describe prehuman evolution with intricate patterns of shared characteristics, rather than a simple 'branching tree'. While it is not yet clear where each one of the 'missing links' fits in the overall picture, Darwin's conjecture – that the ancestors of modern humans arose out of Africa – has been supported by an impressive bank of fossil evidence from the rift valley.

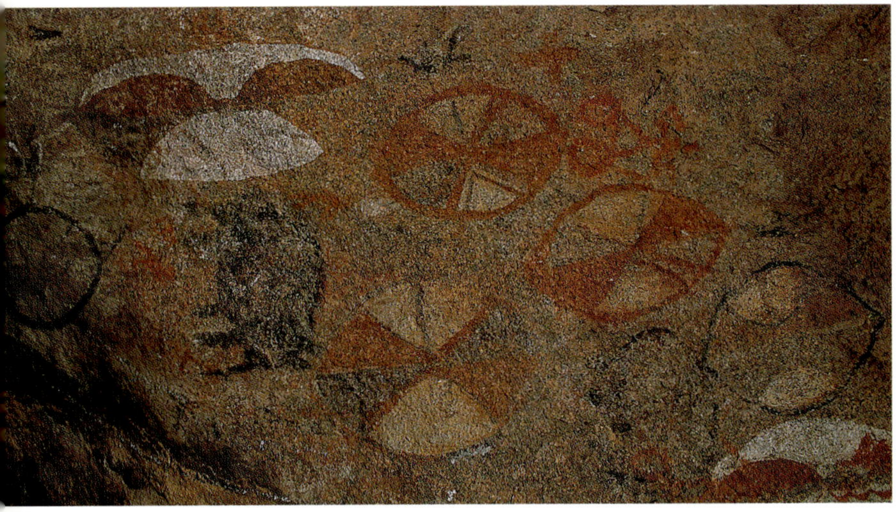

As his name implied, *Homo erectus* walked upright; his brain, though smaller than our own, was larger than that of *Homo habilis*. It probably contained a region of the frontal lobe called Broca's area, associated with speech. *Homo erectus* also had a face more closely resembling that of humans, and may represent an important dividing line in human evolution: everything that came before him was ape-like in character; everything that came after him was human-like. Compared with all that had gone before, the species would have been very human in form as well as behaviour.

Present-day excavations

Today, excavations continue at all the Leakey sites. Two years after Mary Leakey left Olduvai Gorge in 1983, a partial skeleton, attributed to *Homo habilis*, was discovered in Bed 1 deposits by an American team; in 1996, another American-led team found an upper jaw attributed to this species. Three years later, at Lomekwi, 10 kilometres (six miles) west of Lake Turkana, Maeve Leakey's team found the skull of *Kenyanthropus platyops*. Dated at 3.5 million years old, it was thought by many to be another line of 'southern apes' that met a dead end.

A rock shelter at Moru Kopjes in the Serengeti is decorated with painted motifs of Maasai origin.

Elsewhere in the rift valley fossils continue to be unearthed. The most famous is 'Lucy', a 3.2 million-year-old 'southern ape', whose bones were found near Awash River, in the Ethiopian section of the rift valley. In 2000, the bones of a 3.3 million-year-old infant 'southern ape', 'Dikika Child', were found near the same site. Further afield, in Chad, in beds thought to be between six and seven million years old, fossils of yet another species of 'southern ape' were discovered.

Human origins

In short, the human fossil record has become denser. Contrary to the expectations of some early anthropologists, the fossils show that upright walking, or bipedalism, arose before the development of enlarged brains.

There is still no consensus on the subject of human origins. Two core elements of human evolution are, however, undisputed: humans descended from apes that lived in Africa seven million years ago, and fossils of ancestors more remote than *Homo erectus* have been found only in Africa. The details of a more comprehensive account remain speculative, because we do not know exactly how many African migrations occurred over the last four to five million years. The version of events that follows – 'out of Africa' – is the one to which most scientists now subscribe.

About five million years ago, probably in response to a drying of the climate that caused forest cover in Africa to shrink, one great ape species moved into a new ecological niche – the wooded savannah. The descendents of this ape evolved into 'southern apes', such as Nutcracker Man, *Paranthropus Boisei*, of Olduvai Gorge, *Kenyanthropus platyops* of Lake Turkana, and Lucy of Awash Valley. We do not know how many such 'southern ape' lines existed. Between 4.5 and 4 million years ago, the 'southern apes' evolved an upright stance.

'Southern apes' of various species persisted for over three million years before all but one of them died out 1.5 million years ago – the line that evolved into *Homo habilis*, such as Handyman, from Olduvai Gorge. ('Out of Asia' proponents argue that a 'southern ape' species may have migrated out of Africa and evolved into a *Homo* line in Asia.)

Nevertheless, Nutcracker Man brought fame to Olduvai Gorge. The next major find, made by one of Leakey's sons, Jonathan, was assigned to the species *Homo habilis*, or 'Handyman', the species probably responsible for making some of the tools found at Olduvai. Handyman was more ape than human, but his brain size was considerably larger than that of Nutcracker Man. Conventionally, the *Homo* line begins with Handyman, *Homo habilis*, and concludes with humans, *Homo sapiens*; but in between the two, there have been half a dozen or so other *Homo* species.

Lake Turkana

It had not occurred to scientists to excavate the sedimentary deposits bordering Lake Turkana, but a chance occurrence revealed its potential as a productive site for early human remains. In 1969 Richard Leakey, another of Louis Leakey's sons, and Maeve Epps came upon a prehuman skull lying on the surface of a dried-up river bed. The skull belonged to *Homo habilis*, the same species as Handyman of Olduvai Gorge, excavated 10 years before.

Lake Turkana was to yield up further treasures: another *Homo habilis* skull was found. Labelled '1470', it was older than the Olduvai specimen, dating back more than two million years. Then, in 1984, Richard Leakey's team found a nearly complete *Homo erectus* skeleton. They called it 'Turkana Boy', and estimated that it had lived 1.54 million years ago, and died aged between nine and 12 years. It had an entirely modern body structure. Although *Homo erectus* had been known about for almost a century – a Dutch scientist had discovered the first specimen in Java in 1891 – it was known only from scattered fragments; this specimen was an almost complete skeleton.

The site of Japanese palaentological excavations near Lake Turkana in the rift valley, northern Kenya. The search for hominid remains has, to date, delivered limited but encouraging results.

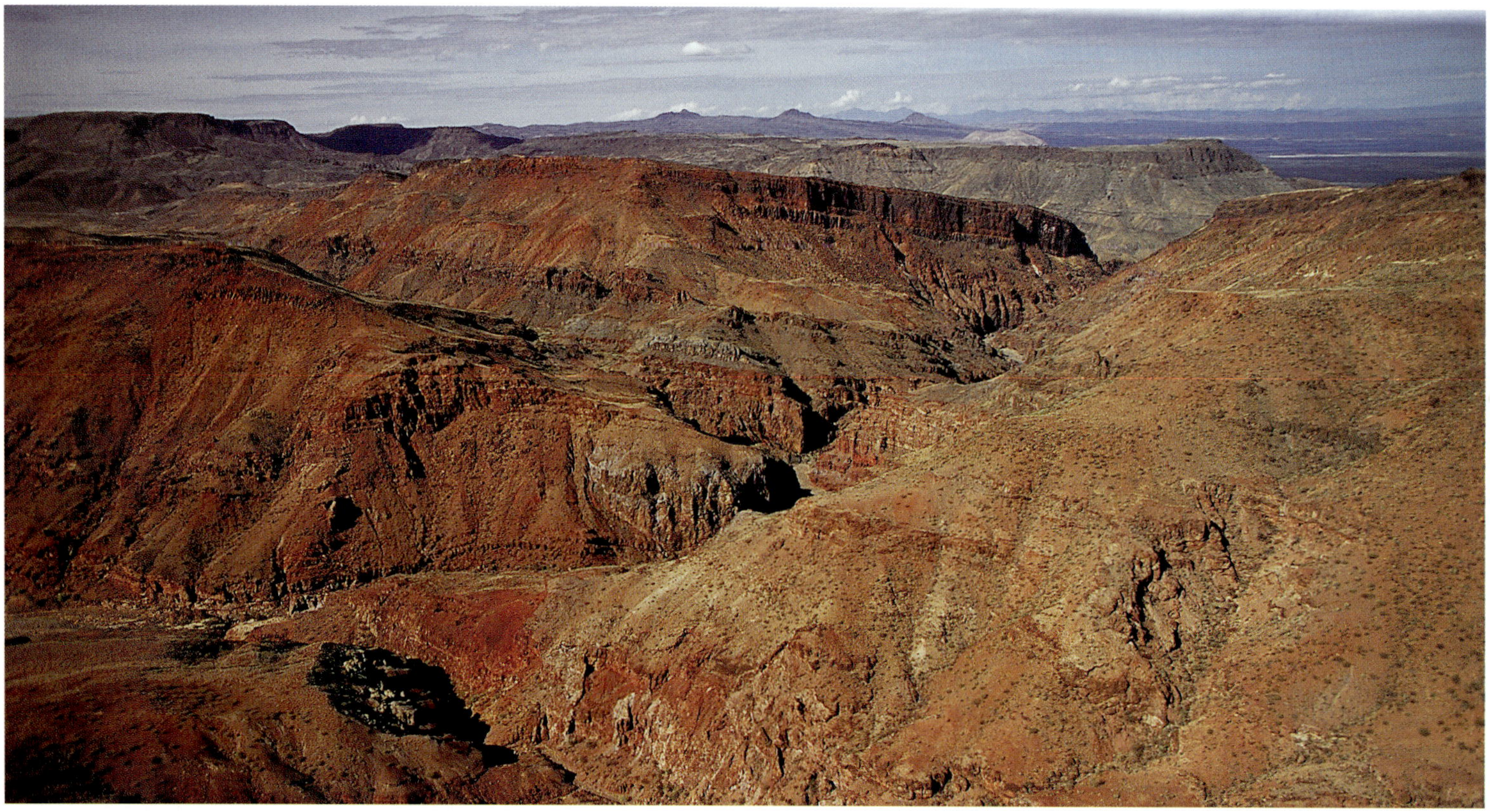

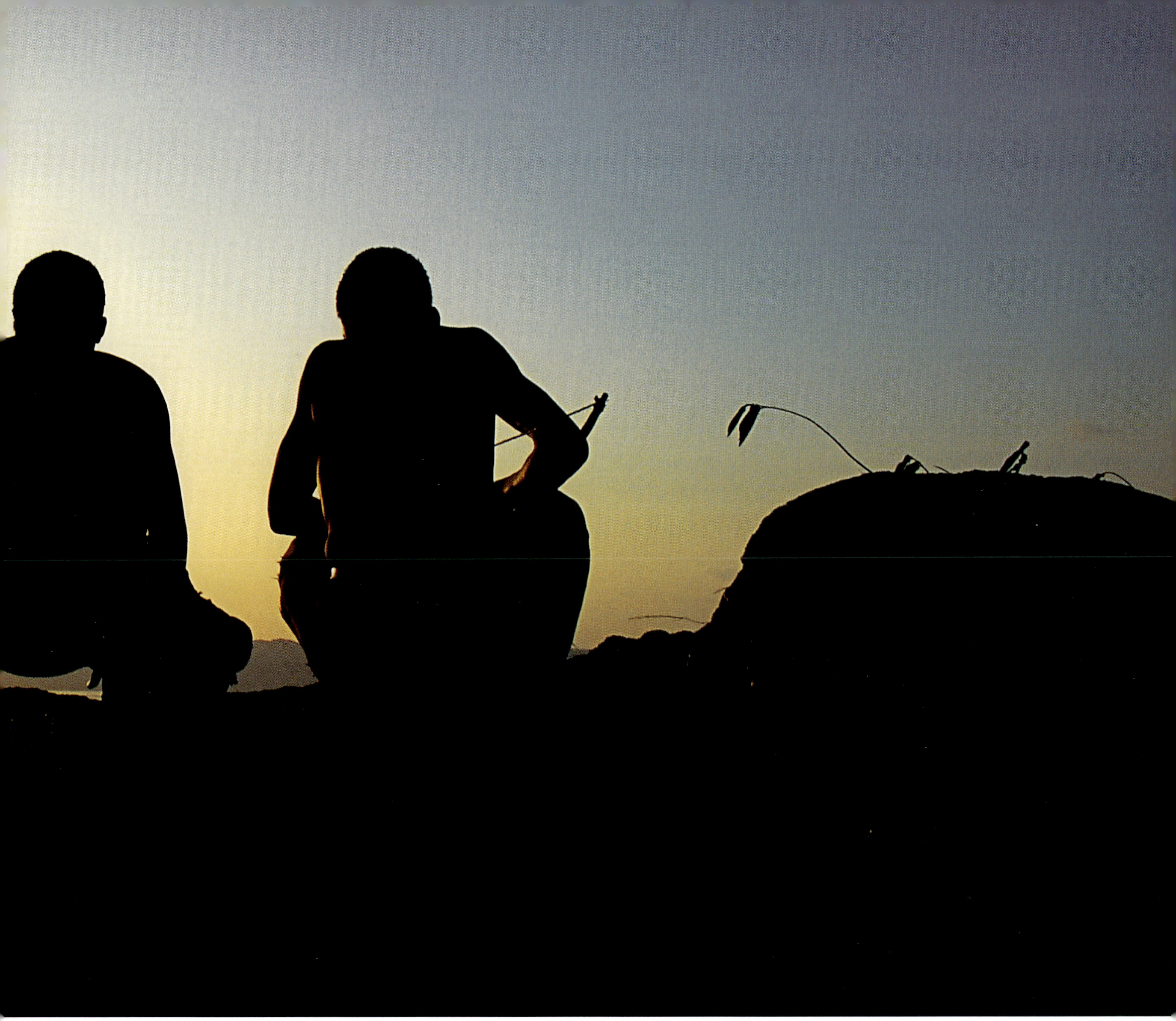

8 Hunter-gatherers of Lake Eyasi

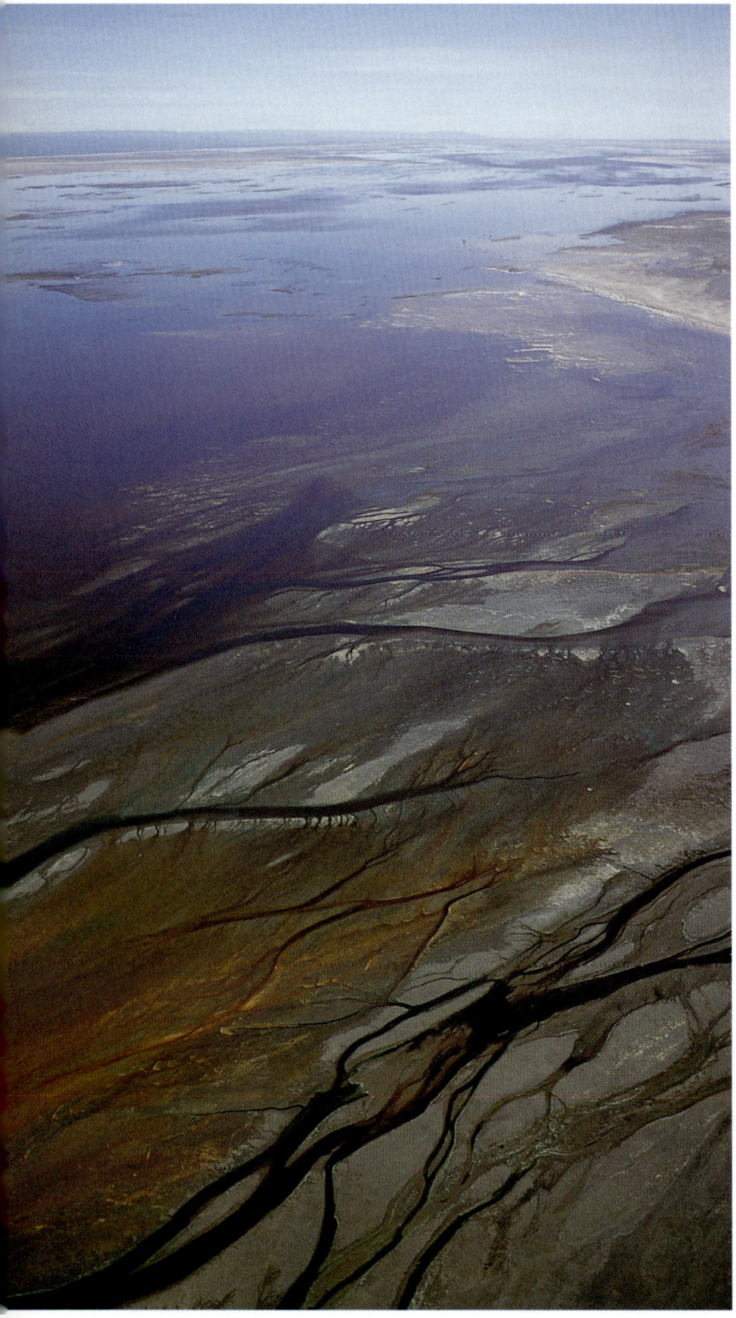

A false arm of the rift valley runs southeast from the Crater Highlands to Lake Eyasi, a shallow alkaline lake that varies considerably in size from one year to the next. The lake basin is the traditional home of the Hadzabe.

The lake basin

At Lake Manyara, situated at the base of the Crater Highlands in northern Tanzania, the main arm of the eastern rift becomes difficult to discern; a little distance further south it becomes distinct again, as it continues southward on its grand and imposing course, leaving behind a short offshoot near Ngorongoro Crater in the Crater Highlands. This short arm of the rift valley features an impressive western wall, 300 metres (984 feet) high, enclosing a shallow alkaline lake – Lake Eyasi. The lake is situated relatively near both Olduvai Gorge and Laetoli, two sites where fossils and evidence of prehuman hominids have been found. On the shore of Lake Eyasi itself, a 130 000-year-old hominid fossil was recently discovered.

It is highly likely that the first *Homo sapiens* – no different, except in the most superficial details, from any modern human – first appeared in this part of the world 100 000 years ago. Today, on the very site where our direct ancestors once lived, a nomadic people, the Hadzabe, follow a way of life – hunting and gathering – that may closely resemble the lifestyle of those ancestors.

The Hadzabe

The Hadzabe number between 1 000 and 1 500 Hadza individuals; their traditional area in Lake Eyasi basin covers approximately 1 500 square kilometres (585 square miles) – fewer than one Hadza per square kilometre (0.39 square miles). Their land is semi-arid savannah, with less than 400 millimetres (16 inches) of rainfall annually, high mean temperatures, and a terrain that is not naturally suited to agriculture – bush land, with many acacias and a few baobab trees. Here, the nomadic Hadzabe have practised a hunter-gatherer lifestyle for centuries, moving their base camps with the rhythm of the seasons.

Until about 20 years ago, the land on which the Hadzabe live also supported a great diversity and a high density of wildlife species: buffalo, zebra, giraffe, eland, kudu, vervet monkey, baboon, impala, dikdik and other smaller animals. Thirty years ago, they shared their land with herds of as many as 200 elephant; a decade before that, the Hadzabe routinely hunted rhino. Today, hunting is more difficult, given the depleted herds, and the men often slip into the neighboring Ngorongoro Conservation Area, where the wildlife is more prolific. The men also gather three types of honey from wild sources; honey gathering is seasonal since the water and the flowers that are essential to wild bees are also seasonal. Women gather foods – berries, fruits and a variety of tubers – from wild plants. Rather than relying on one staple crop or product such as maize or livestock, the Hadzabe subsist on a wide variety of plant foods and wild animals, avoiding only the spotted hyaena. While many of these resources are available only seasonally, they are sufficiently diverse to accommodate periodic shortages, with the result that the Hadzabe diet is unusually varied for subSaharan Africa.

A mobile society

Hadza society is described as being 'fission-fusion': the nucleus is the family unit, to which relatives attach and detach; the family group is dynamic, and so the society is fluid. The predominant feature of Hadza life is mobility; because some of their food is mobile and some of it is seasonal, families move from camp to camp to avoid depleting food resources in any particular area; such efficiency is central to a hunter-gatherer lifestyle.

One consequence of mobility is that the Hadzabe have very few possessions – a natural outcome of the basic incompatibility between mobility and materialism. As a result their culture is recorded in their memories, not carried on their backs. It is transmitted orally by singing and story-telling.

In keeping with a mobile lifestyle, mothers space out the births of their children over several years. Mobility is physically demanding; mothers must carry their infants as they perform their tasks – gathering food, or migrating with the group from an old camp to a new location. Infants continue to suckle until up to four years of age, which may have the effect of making it harder for a woman to conceive. By the time she does give birth to another baby, the previous child has reached a level of independence, and is able not only to do without suckling, but also to walk far and fast.

A small party of Hadza hunters returns to base camp from a hunting foray.

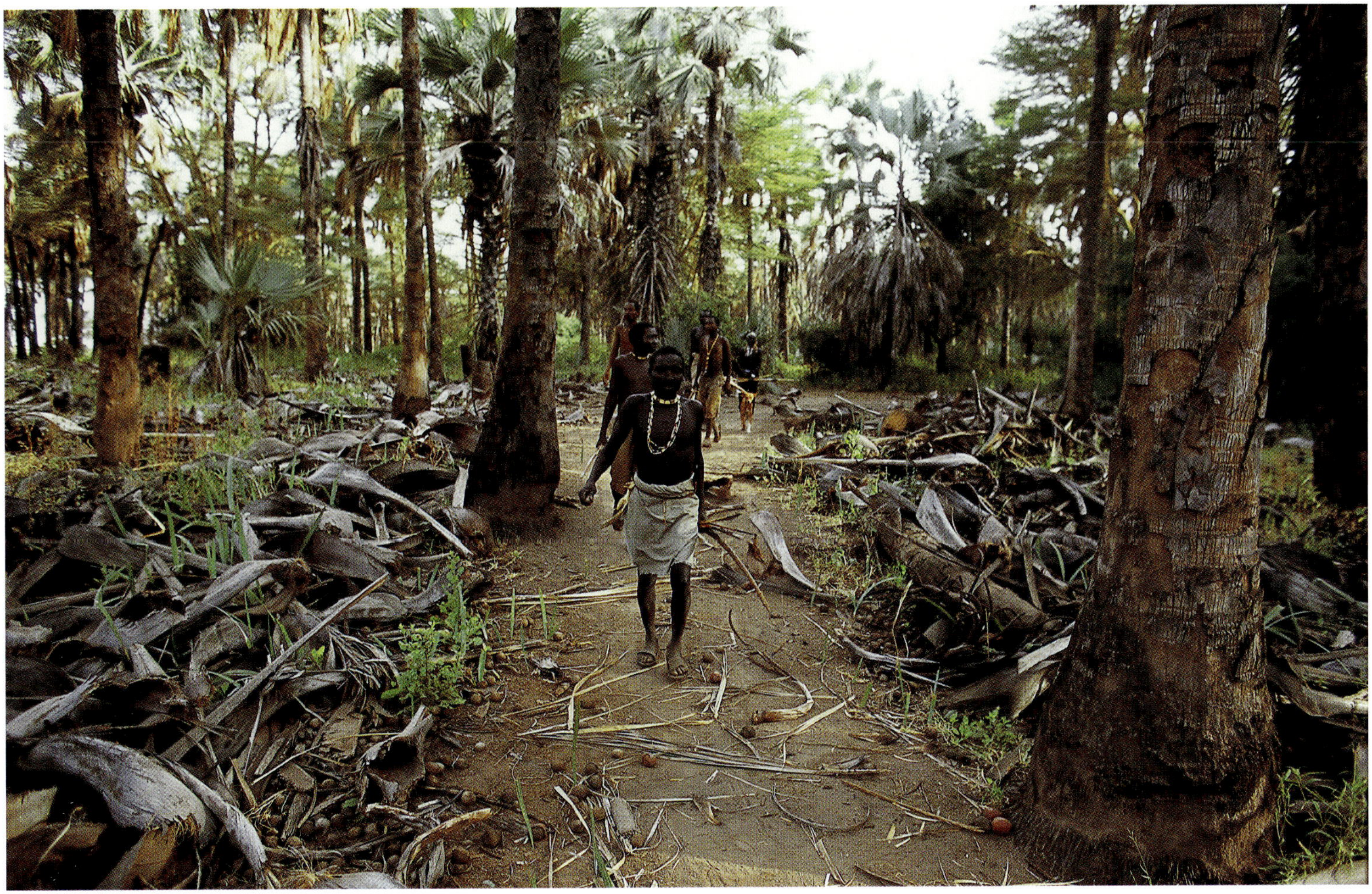

Opposite: *The bond between mother and child is a close one. Today, by law, Hadza children are required to attend school, but many of those sent to the nearest boarding school run away to rejoin their parents.*
Below: *Target practice: the men have good control, but their handmade bows and arrows are less accurate than those made with the aid of precision tools.*

Division of labour

Hadza men hunt for meat and gather honey; the women collect plant foods. This division of labour according to gender appears to be based on infant care, especially nursing, which is best provided for by mothers; additionally, tracking and stalking animals requires silence, and noisy infants would be a handicap, were mothers to hunt.

Gathering edible plants, on the other hand, is a highly social activity, involving several mothers and their older children. It is also a skilled occupation: a Hadza woman needs to know where suitable plants are to be found, in which season they are edible, and how to keep oriented in the bush. The only technology employed is a suitably shaped branch for extracting tubers.

Hunting is a much more solitary affair, requiring silence and stealth. The Hadza hunter uses simple technology – primarily one-and-a-half-metre-long (five foot) bows, and two types of arrow, fashioned from the thin branches of selected bushes and small trees, along with feathers plucked from guineafowl and vulture. The arrows used to kill birds and small game are simply sharpened to a point, but those selected for large game are tipped with metal barbs and dipped in a fast-acting poison, made by boiling the sap of the desert rose tree, *Adenium obesum*, which grows

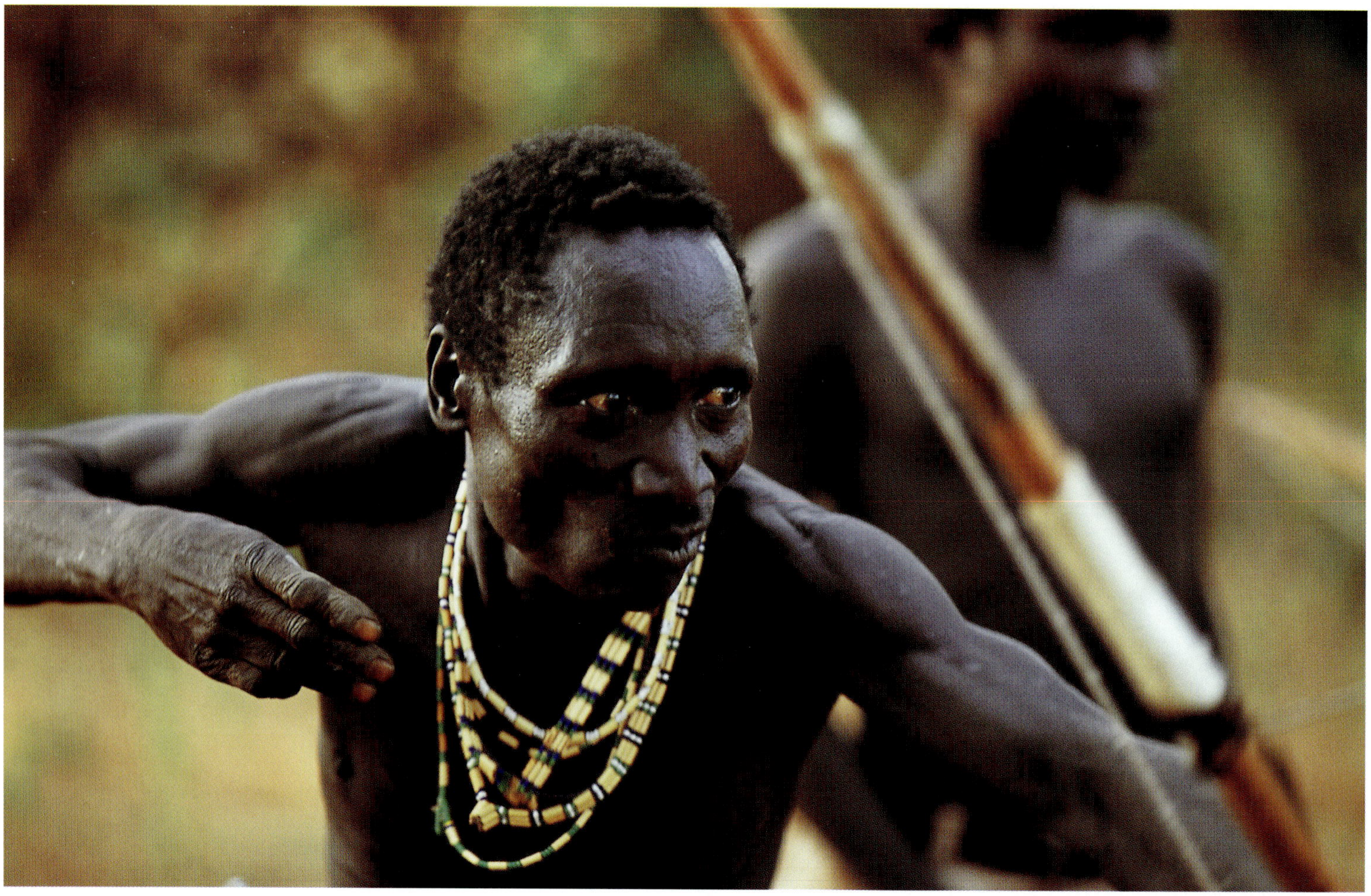

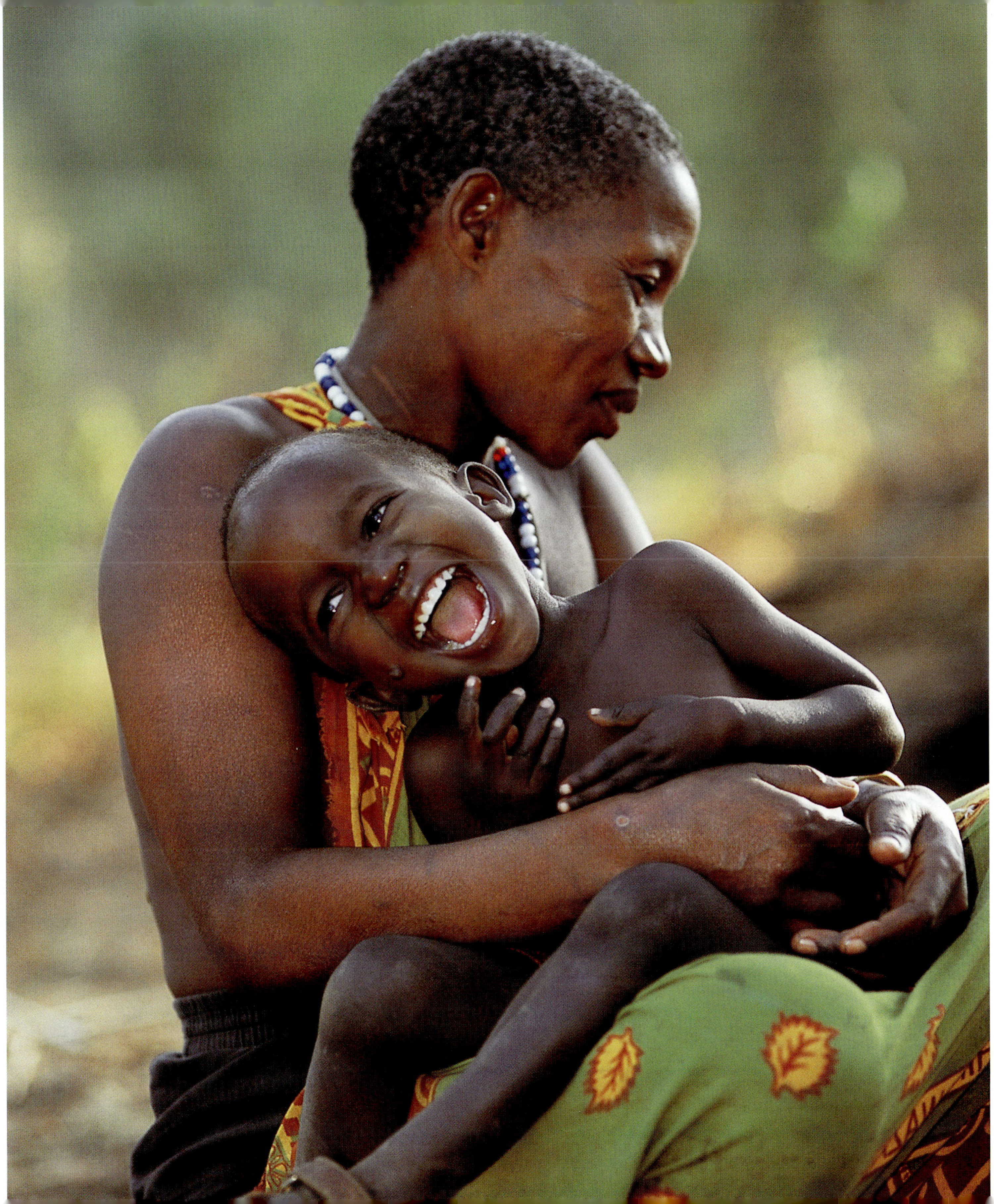

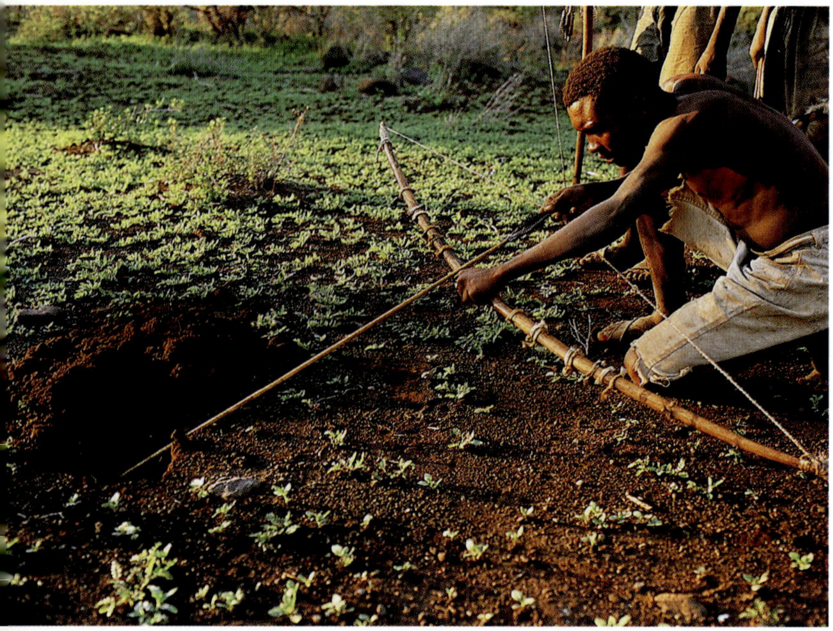

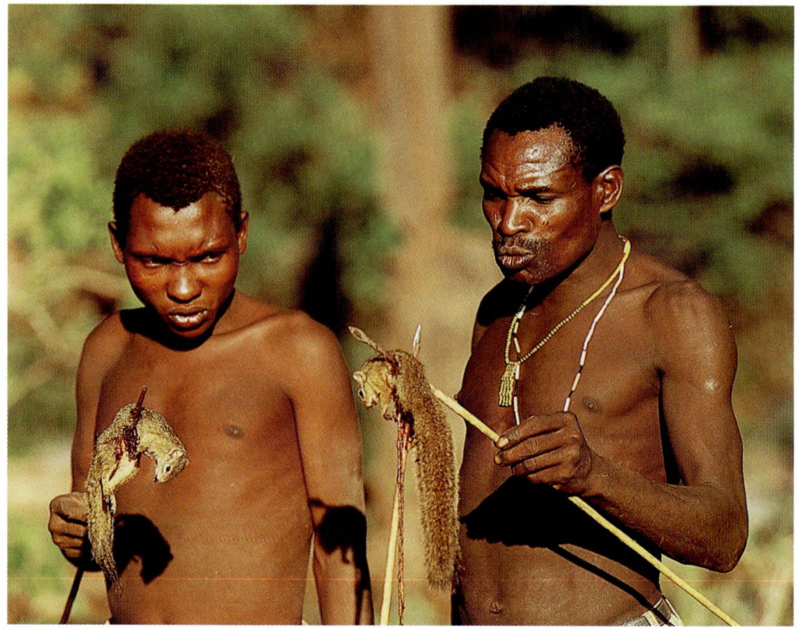

Above: *A small mammal escapes into a burrow at the approach of a Hadza hunting party; a hunter attempts to shoot it in the burrow, but without success.*
Above right: *In the past, rewards from hunting have been small, relative to the effort involved, but more recently they have been meagre – a squirrel or two shot at close range constitutes a morning's work.*

defiantly in a few locations despite the poor soil. After the sap has been boiled thoroughly, a sticky residue remains. It is safe to the touch, but lethal on entering an animal's bloodstream. It can knock out a buffalo in minutes.

Although the technology used in hunting is simple, the skills employed – which have evolved over many thousands of years – are considerable. The hunters have an uncanny ability to track animals over difficult terrain. From an animal's tracks a Hadza hunter can identify not only the animal's species, but also its age and sex, how long ago it left its tracks, whether it was running or moving slowly, and whether it was injured or fully fit. Once within striking distance of the victim, the hunter uses considerable cunning to stalk it. With the animal in sight, he seems to know intuitively how it will react, and displays unerring tactical co-ordination in closing in on it.

An egalitarian, peaceful, sustainable lifestyle

By its very nature, hunting success is highly variable, whereas the gathering of plant food is more certain, and also more productive; in consequence, the Hadza diet leans more toward plant foods, which form a larger proportion of the diet than do meat and honey. Because plant gathering is more time-consuming than hunting, women work longer hours than men, but men and women are treated as nothing less than equals in Hadza society. The egalitarian nature of the Hadza way of life applies in a broader sense than just gender equality. There is no hierarchy: no leaders in their ranks, no wise elders guiding the community and settling disputes. It is truly a 'bottom-up' society as opposed to a 'top-down' one.

The Hadzabe do not interact confrontationally. A fortuitous outcome of social fluidity – camps changing location, size and composition regularly – is the avoidance of conflict. If there are disagreements, parties in dispute simply move away from one another in the course of normal life. The easy-going Hadzabe appear to have few problems involving material possessions and interpersonal encounters, since they own very little, and are relatively thinly distributed across the land.

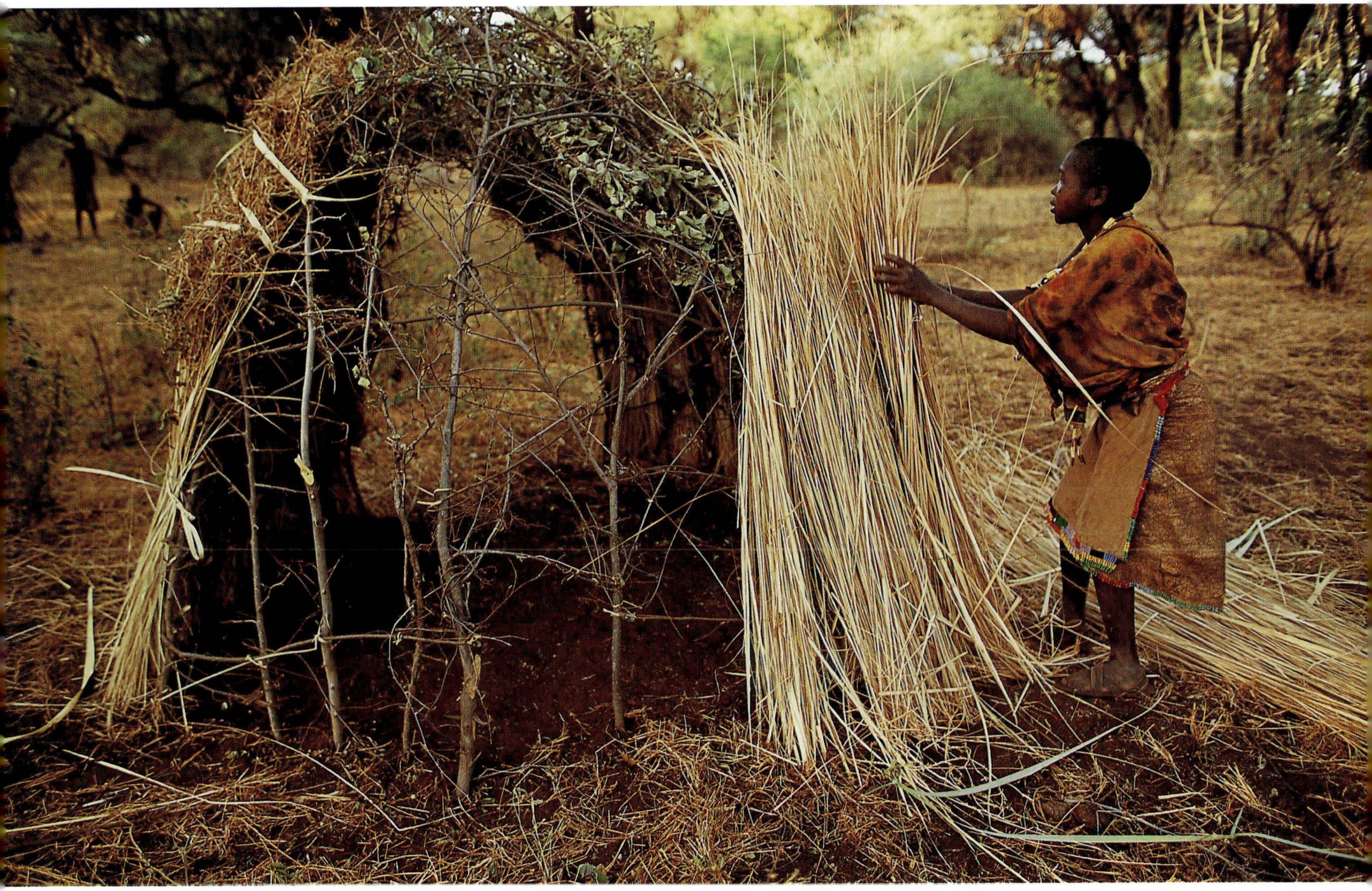

Hadza women stay behind and set up a temporary camp while the men go hunting.

The Hadzabe have an 'immediate-return' economy: present needs are met. Resources are normally consumed on the day they are gathered. The Hadzabe do not have set meal times; they eat when they find food. When a large animal is killed, meat is not stored; it is shared out immediately with neighbouring camps. This way of life contrasts with a 'delayed-return' economy – that of herders and farmers – in which current and future needs are provided for.

The secret of successful day-to-day living seems to lie in having a secure and intact land base in the wilderness, combined with the knowledge and skill required to exploit it sustainably. Even though the Hadzabe live in a part of the rift valley that experiences frequent droughts and unpredictable rainfall, they have never known the famine or hunger experienced by their neighbours, who are dependent on livestock and farming. Furthermore, their skill in both hunting and gathering gives them a far more varied diet than that of their more specialised neighbours.

Their hunter-gatherer lifestyle is not an incessant search for food; a great deal of leisure time is at hand, and this is largely spent in socialising. Hadza sociality extends beyond the confines of the family. Families, scattered in camps, come together at special occasions, such as marriages, and also when there is a surplus of meat. The Hadzabe love eating meat, and no invitations to do so are declined. The sharing of meat runs along lines of kinship, alliances and obligations.

Sharing is a core value among the Hadzabe – within a family, between families, and right up to the boundary of their social universe. Just as modern, free-market economies are characterised by the practice of trade and of owning private property, so the Hadzabe are defined by a belief in common property, sharing and gift-giving.

Human heritage preserved

For the last two million years, our ancestors in the rift valley, from *Homo habilis* onward, have followed a simple life. They have lived within the constraints imposed by the environment; they have had barely any impact on the ecology of their natural home. For a million years they subsisted by scavenging and collecting plant foods. Then

Lean and athletic Hadza hunters run in pursuit of their prey.

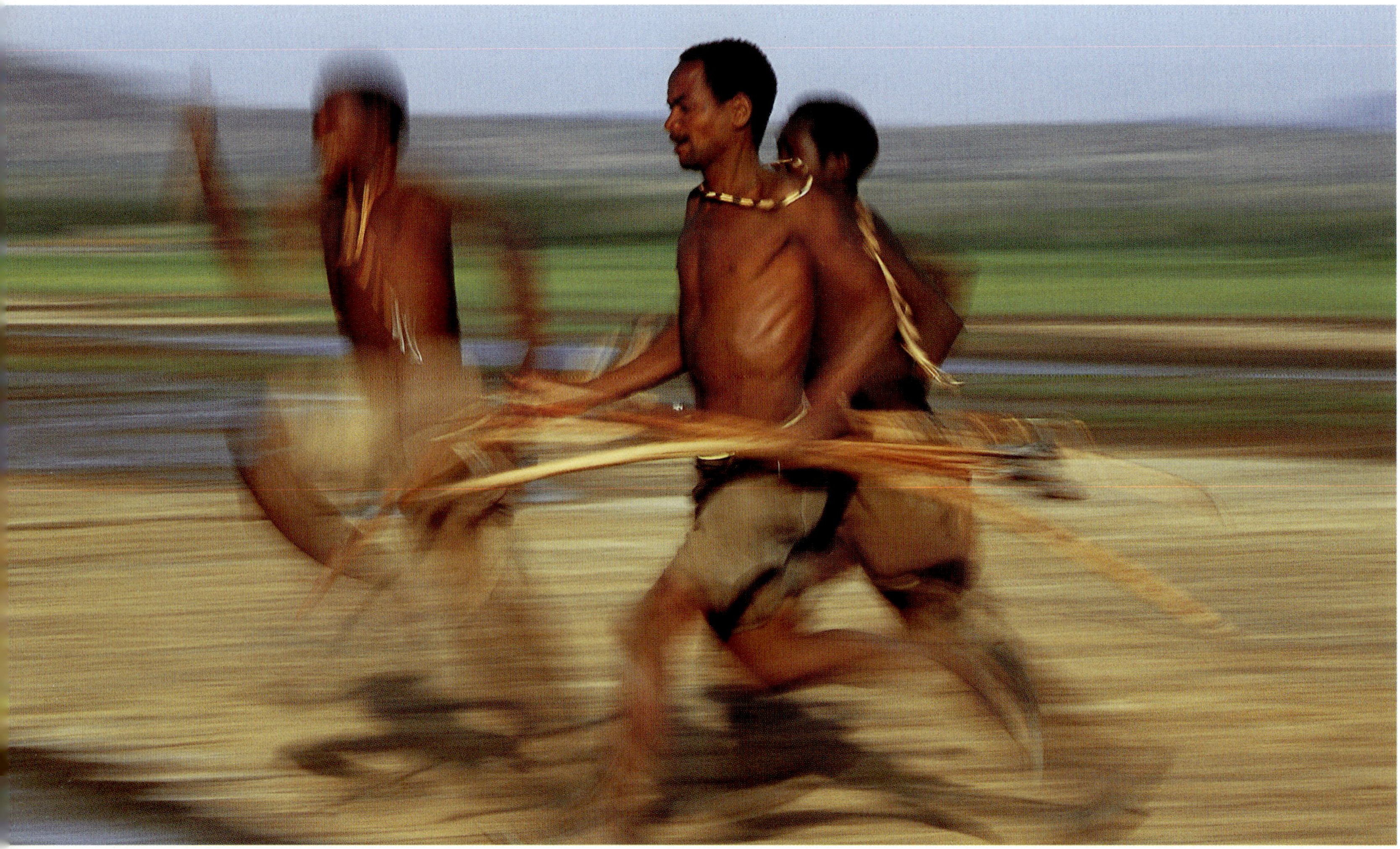

their lifestyle evolved, until hunting and gathering was practised by *Homo sapiens* in Africa. Fossils are of limited use in throwing light on the nature of the lives of early humans. The modern Hadzabe themselves are the best guide we have. Since Hadza society conforms to a 'bottom-up' as opposed to a 'top-down' model, it is likely that our human ancestors functioned in the same way. Like the Hadzabe, they would have valued freedom of movement and personal independence. There is likely to have been division of labour, and a culture of sharing (especially of meat) and of socialising; land and its associated resources would have been common property; and our ancestors must have developed a skilled reading of the environment and the use of simple technology. Their way of life would have left the landscape intact – a concept quite alien to our modern lifestyle.

The Hadzabe have followed much the same lifestyle our ancestors are likely to have practised, in the very place where early human culture first arose. Until only 40 to 20 years ago, they were able to avoid contact with the outside world: the rift valley had provided a sanctuary in which their society was able to evolve, free from the overwhelming influence of outside forces.

Gathered around the fire in the early evening, a Hadza extended family listens to stories and sings.

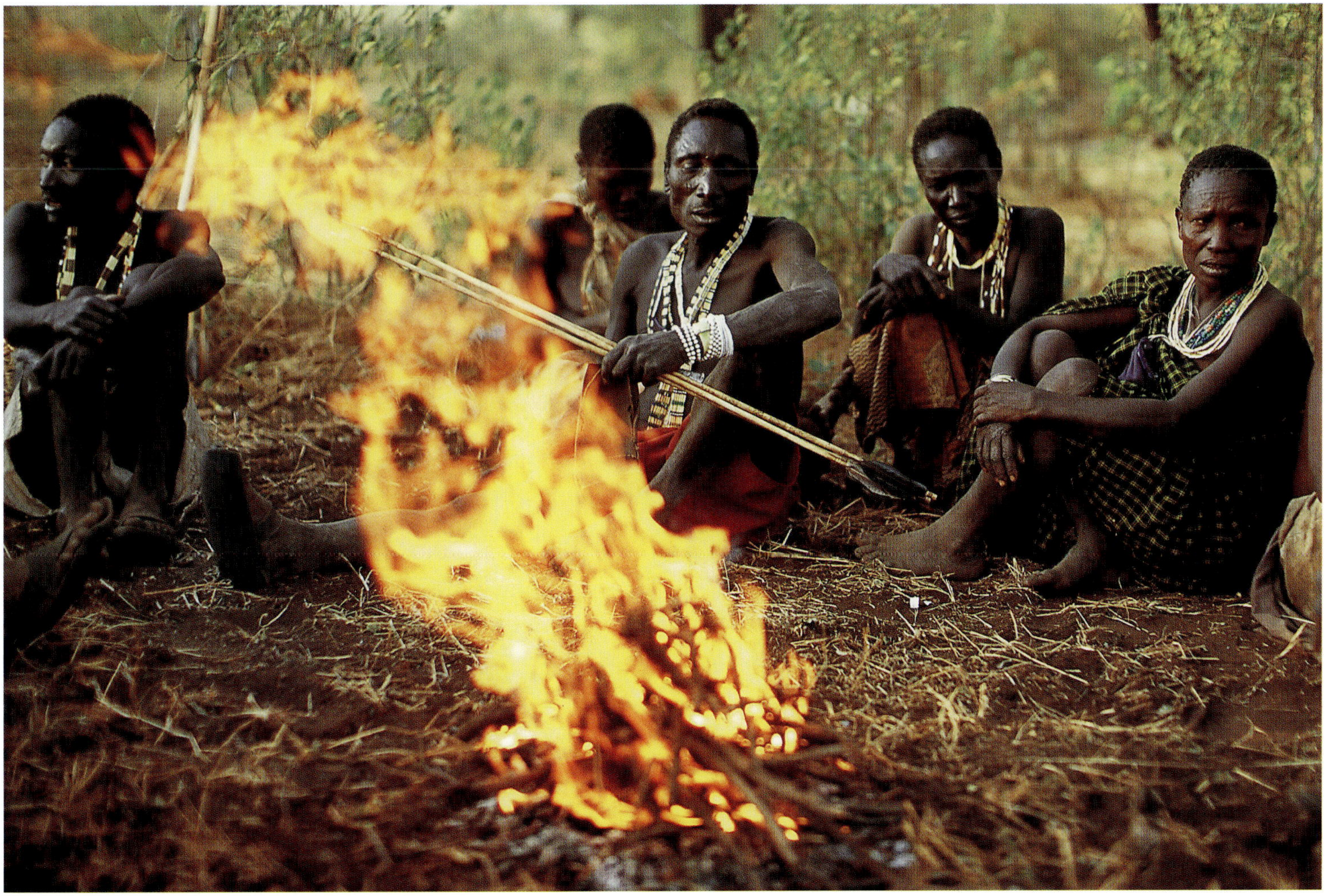

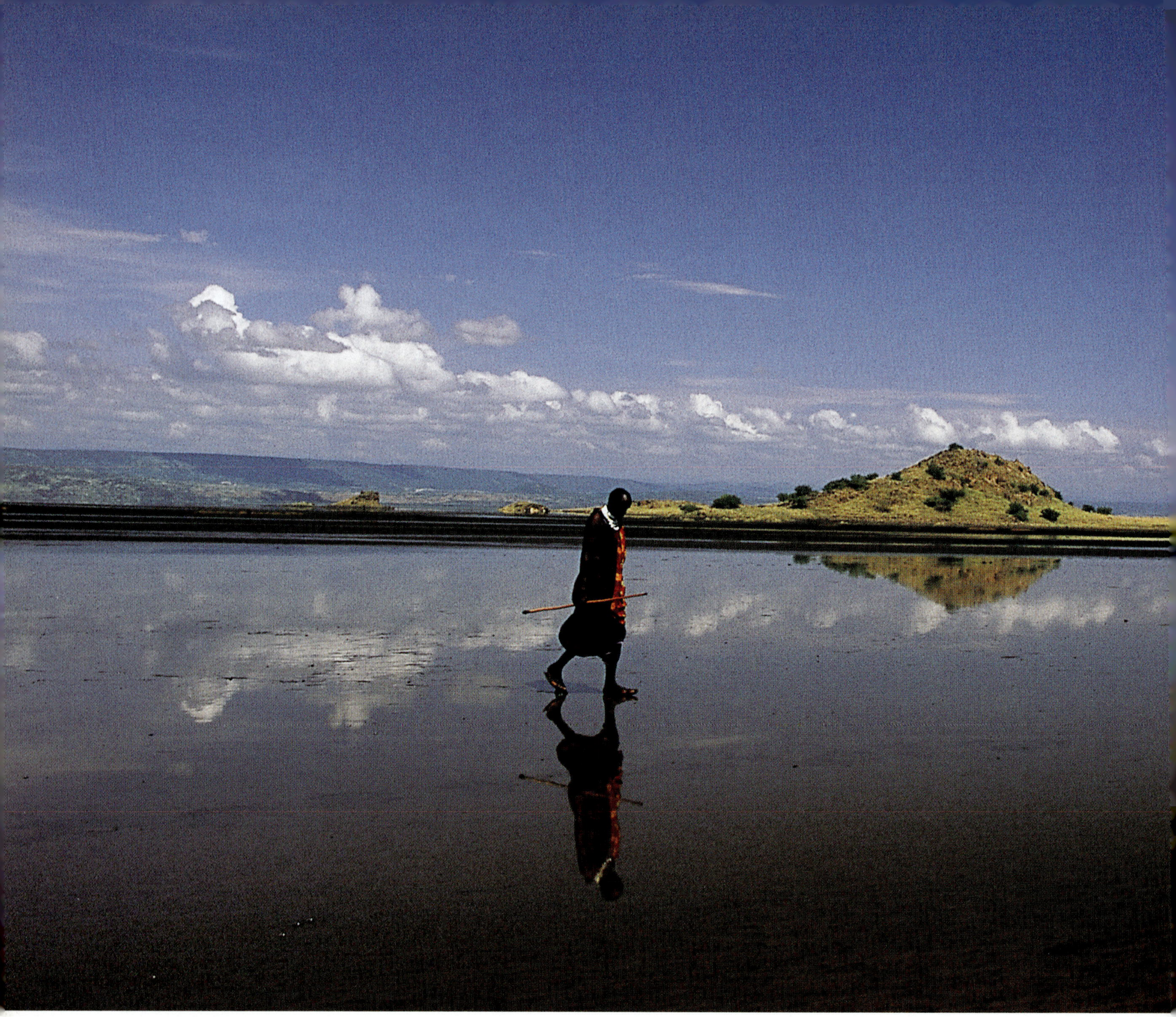

A Maasai man walks along the shore of Lake Natron, looking pensive as he traverses the wet mud.

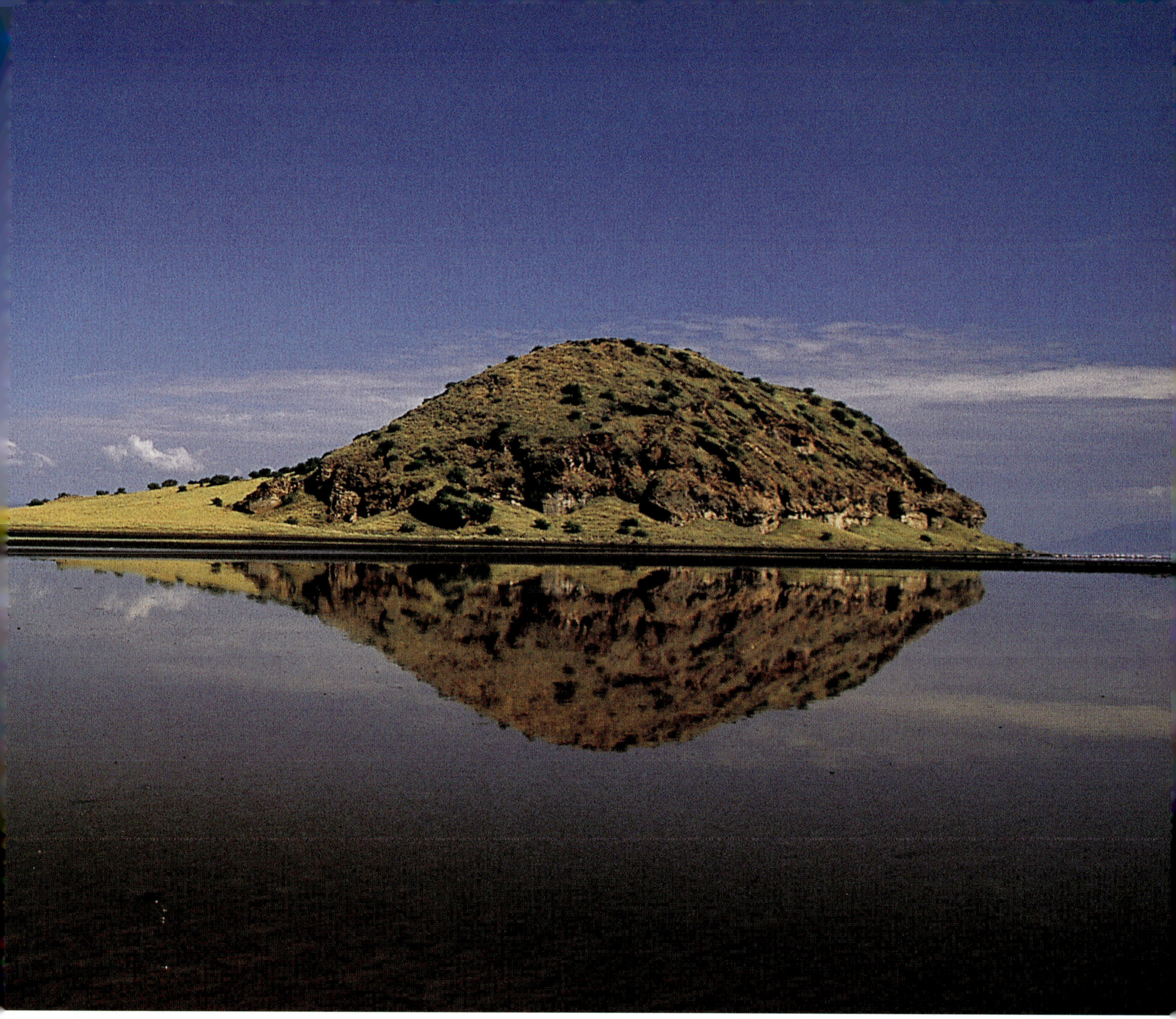

9 Transformation

There are many niches in the rift valley to which various animal species and human groups have adapted perfectly; but their habitats, their way of life – indeed their very survival – are now threatened.

Between 80 000 and 50 000 years ago, a species had evolved with hands and big brains, walking upright – and with a prodigious ability to travel. These generalists migrated to other parts of Africa, crossed over to Asia and Europe, multiplied, farmed, formed complex societies, devised institutions of religion, government, justice, army and police, and used their intense concentration and insatiable curiosity to make scientific discoveries and to build machines. Throughout this drawn-out process, specialists like the pastoralist Maasai and the hunter-gatherer Hadzabe continued to live in the rift valley, in harmony with nature, having little permanent impact on their surroundings.

A Maasai 'boma' on the slopes of the Crater Highlands. With the spread of modern medicine, infant mortality has dropped, the population has grown and settlements have proliferated.

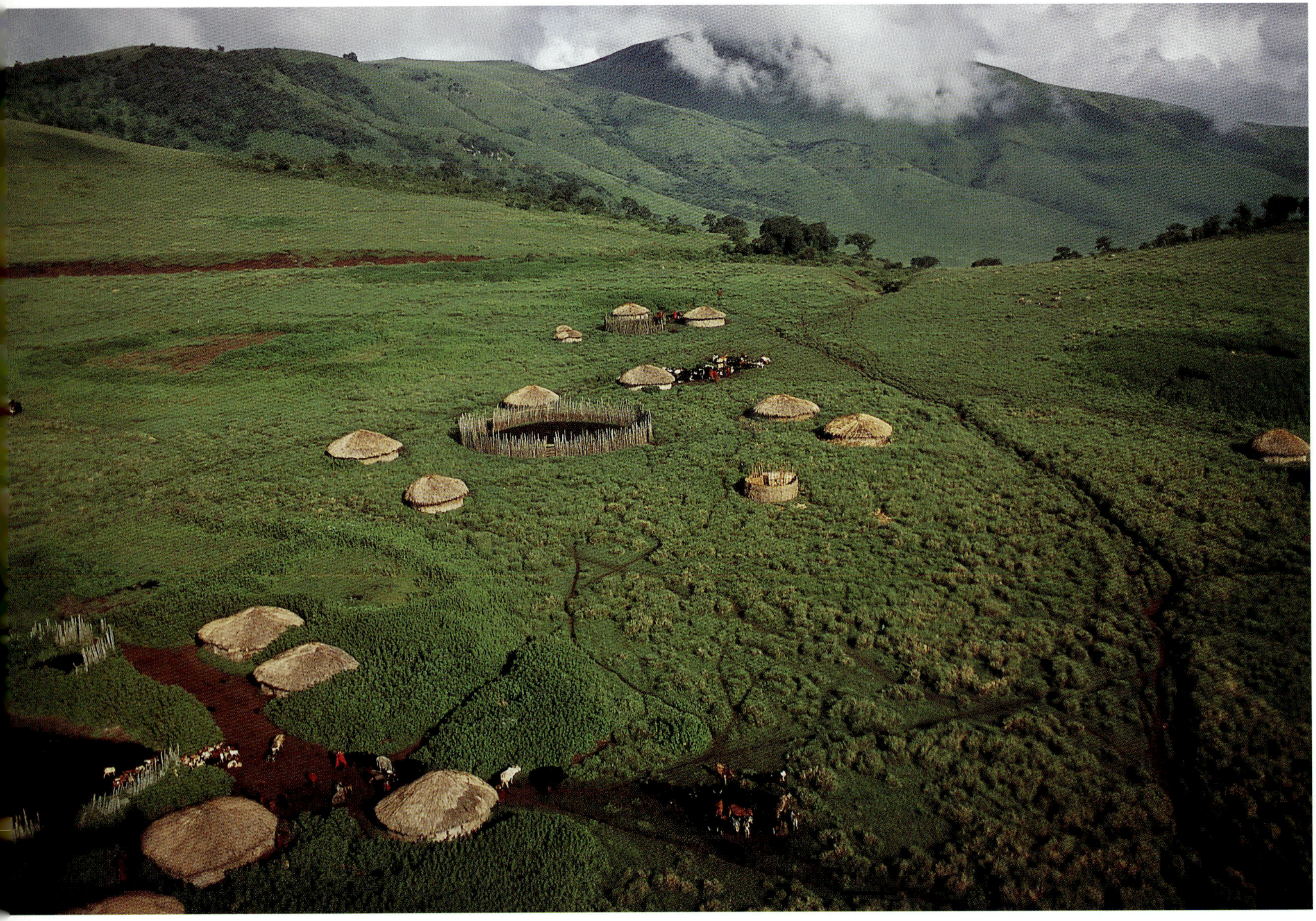

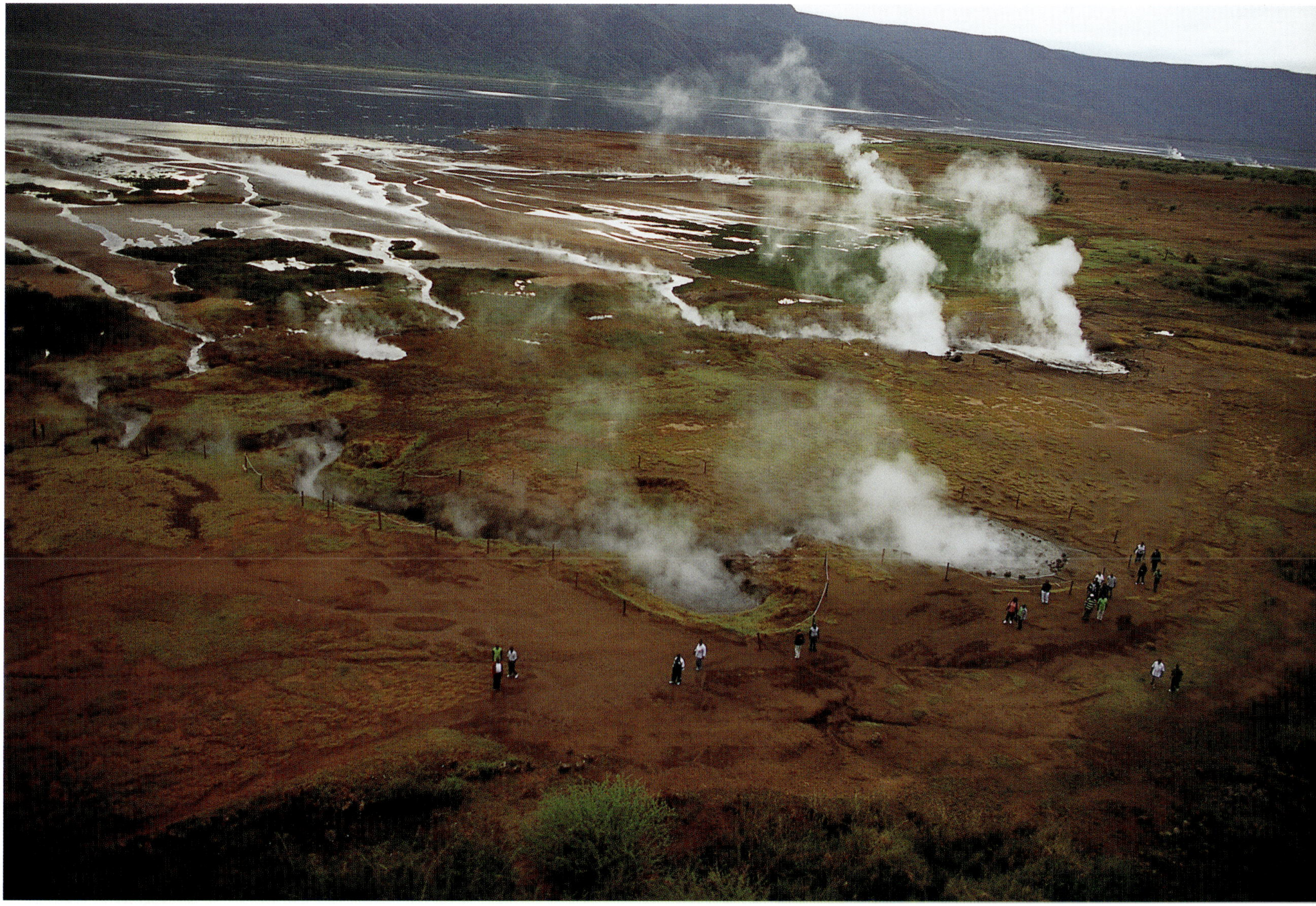

Generalists and specialists

Tourist pressure at the hot-water geysers at Lake Bogoria, where flamingos come to drink fresh water, has led to the area's being cordoned off.

In recent centuries, the Maasai have spread through many parts of the rift valley. They practise pastoralism, which is predation on an entirely different level: their prey animals live with them. The Maasai take advantage of the animals' strong herd instincts to control them; the animals follow wherever their herders take them, and are available at all times to be milked or killed. The Maasai and their cattle are continually on the roam, in search of grass nourished by the seasonal rains and the rift's volcanic soil.

But during the course of the last century, other humans have entrenched themselves in the rift valley in a different way. Unlike the Maasai, they are generalists. The invasive, technologically superior power wielded by the generalists threatens every level of life that has previously flourished in the valley.

Wolves

Humans are steadily invading the habitat of the Ethiopian wolf; the distinction of being the world's rarest canid (see Chapter 1) has not ensured the protection of this highland species. Ethiopia's wondrous Afro-Alpine highland islands provide run-off water for millions of people in the lowlands of northeast Africa. They also harbour most of Ethiopia's endemic animals and plants. Nevertheless, they continue to be exploited at an unsustainable rate. Even in the two Afro-Alpine national parks – Bale Mountains and Simien Mountains – protection is minimal.

The people who inhabit the Afro-Alpine highlands are accompanied by livestock and domestic dogs. Dogs pose two threats to the highly endangered Ethiopian wolf. Male dogs take advantage of the promiscuity of alpha female wolves, and wolf-dog hybrids, which are fertile, result. This dilutes the purity of the Ethiopian wolf population. Domestic dogs also bring disease with them; rabies has struck Ethiopian wolf populations in Bale Mountains National Park twice in the past 17 years. Between 1991 and 1992, the Bale Mountains population was drastically reduced by a combination of rabies and canine distemper brought in by dogs. In September 2003, a second outbreak of rabies killed between 70 and 80 per cent of the Web Valley population. The long-term future of the Ethiopian wolf depends on a withdrawal by the Ethiopian people from the wolf's remaining habitats – currently a highly unlikely prospect.

An Ethiopian researcher carries out an autopsy on an Ethiopian wolf suspected of having been infected with rabies by domestic dogs.

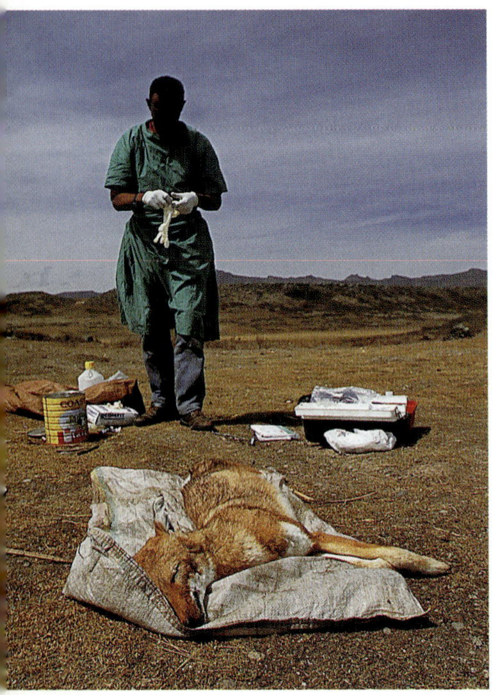

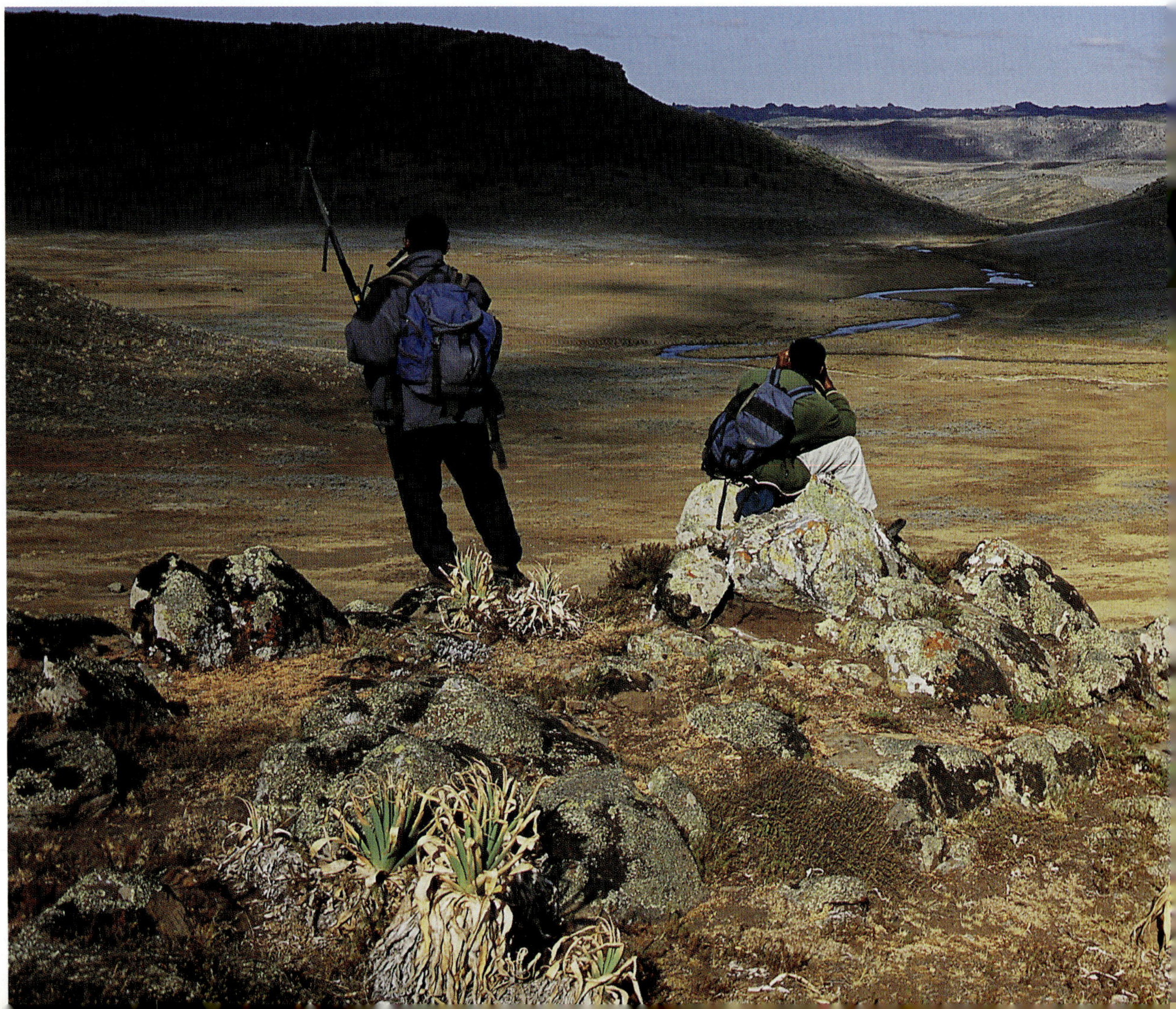

Flamingos

The harsh, hostile world of the soda lakes is home to that extreme specialist, the flamingo (see Chapter 2). It would be hard to find two places less suited to human habitation than Lake Magadi and Lake Natron – but the power of human technology has turned nature into natural resources. Caustic Lake Magadi is the site of a soda factory, and Lake Natron – a critical flamingo breeding site – is next in line; the crucial salinity balance of Natron is also under threat from increased logging and a planned hydroelectric power plant in the lake's main catchment area. Lake Nakuru, also frequented by flamingos, is polluted, bearing the brunt of the sewage from Nakuru town. Furthermore, the rivers feeding the lake are running dry, as deforestation and diversion of water from their sources on the Mau Escarpment proceed unchecked. The march of civilisation, driven by human greed, has contributed to the destruction of the freshwater Lake Naivasha, once a scenic habitat for wildlife on the rift valley floor, now choking with pesticide residues and invasive species, notably water hyacinth. Its tributaries, the Karate stream and the Malewa River, can no longer meet the demand for lake water imposed by agri-business. Charcoal burning, quarry mining and the ubiquitous goats have all contributed to the collapse of this ecosystem. To reverse this trend in ecological degradation, it would be necessary for the major developments in the area to withdraw, incurring huge financial sacrifice.

Ethiopian researchers attempt to pick up radio signals from a radio-collared wolf.

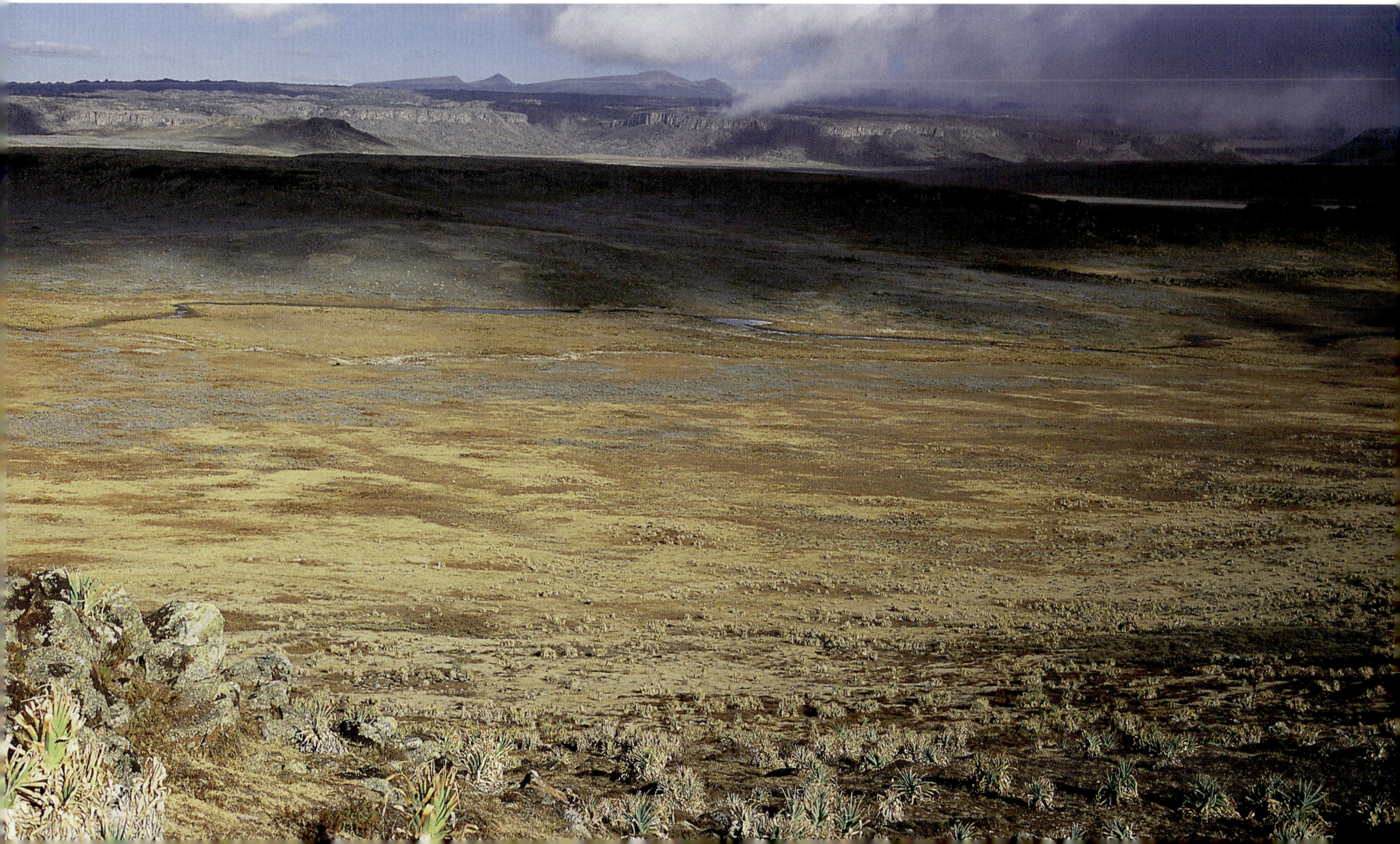

Right: Frankfurt Zoological Society has donated equipment, including cars, radios and binoculars, for monitoring Ngorongoro's black rhino population.
Below: An anti-poaching patrol in the Western Corridor of the Serengeti closes in on a poachers' lair.

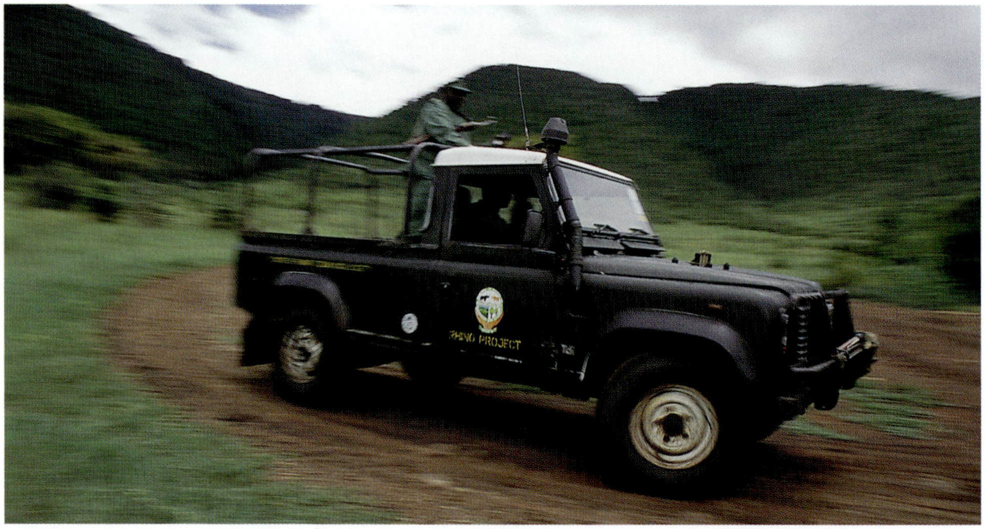

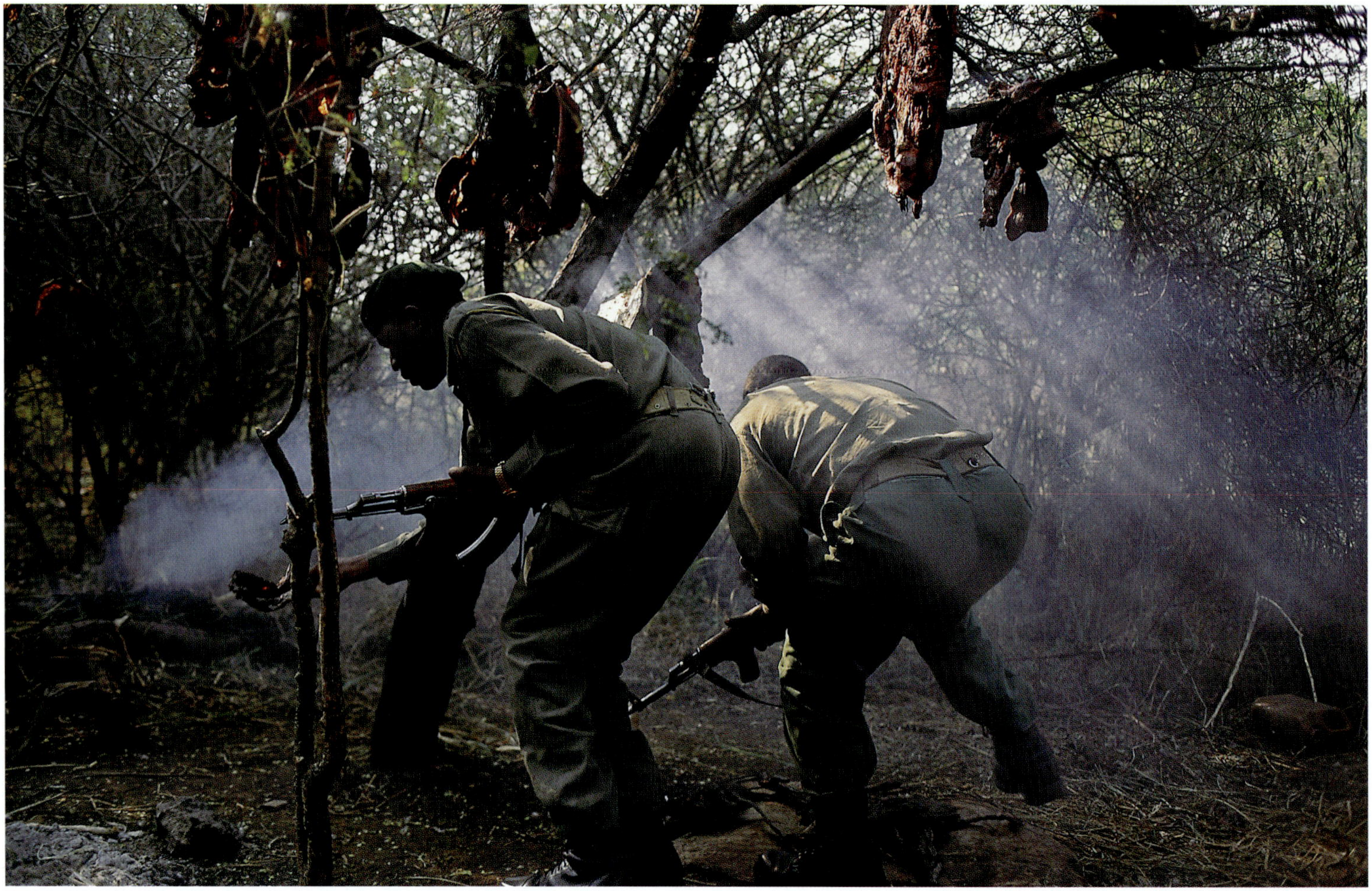

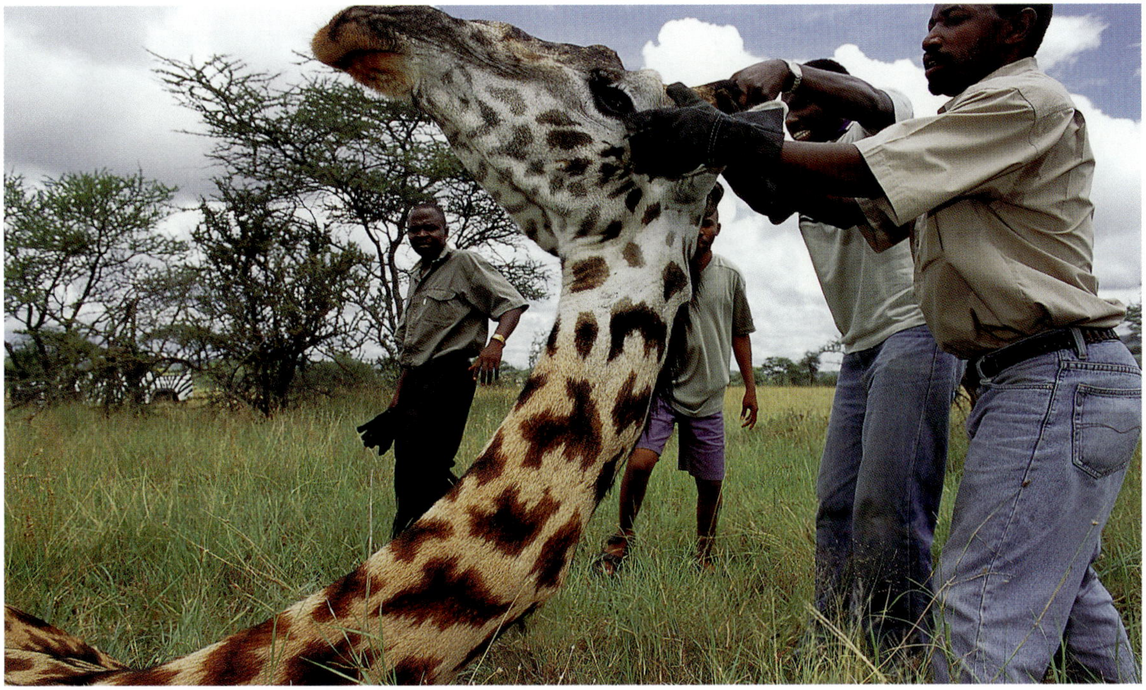

Left: *The veterinary team in the Serengeti in action, desnaring a Maasai giraffe. Once a snared animal is reported, the team tracks it down and darts the animal with anaesthetic; it then removes the snare and collects data on the animal's physiological condition.*
Below: *Workers remove non-indigenous plants, fast encroaching on the Ngorongoro Crater habitat.*

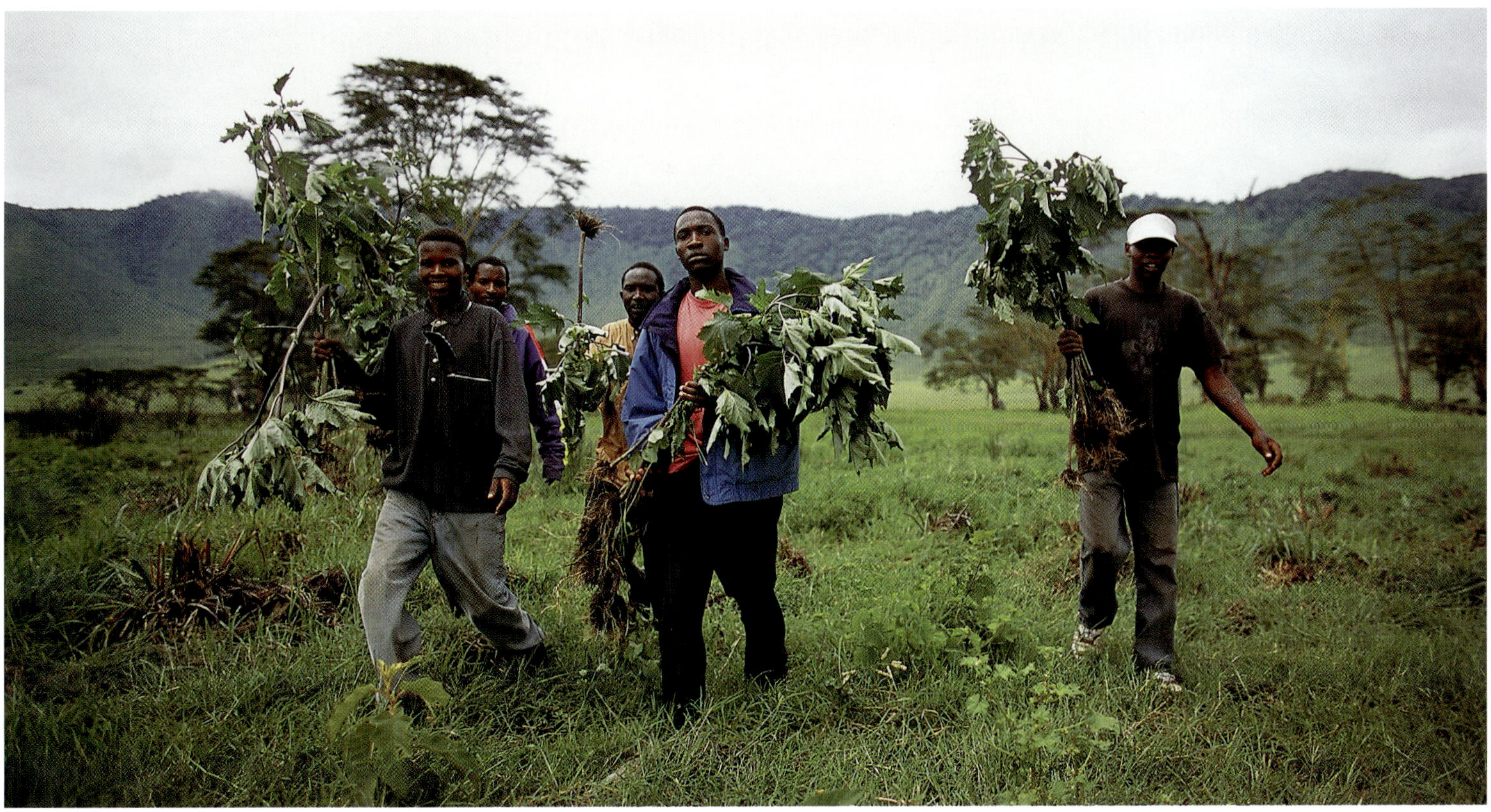

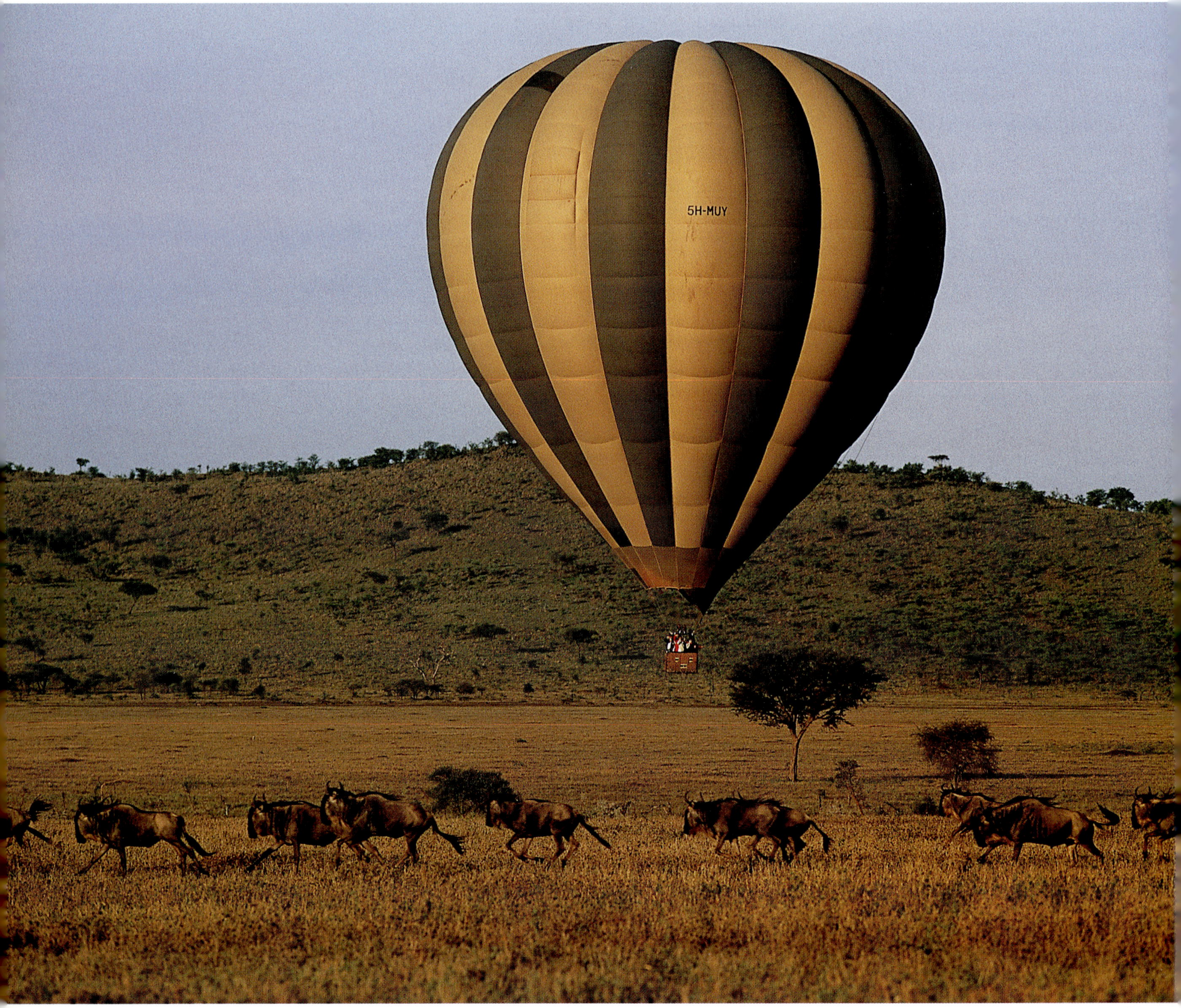
5H-MUY

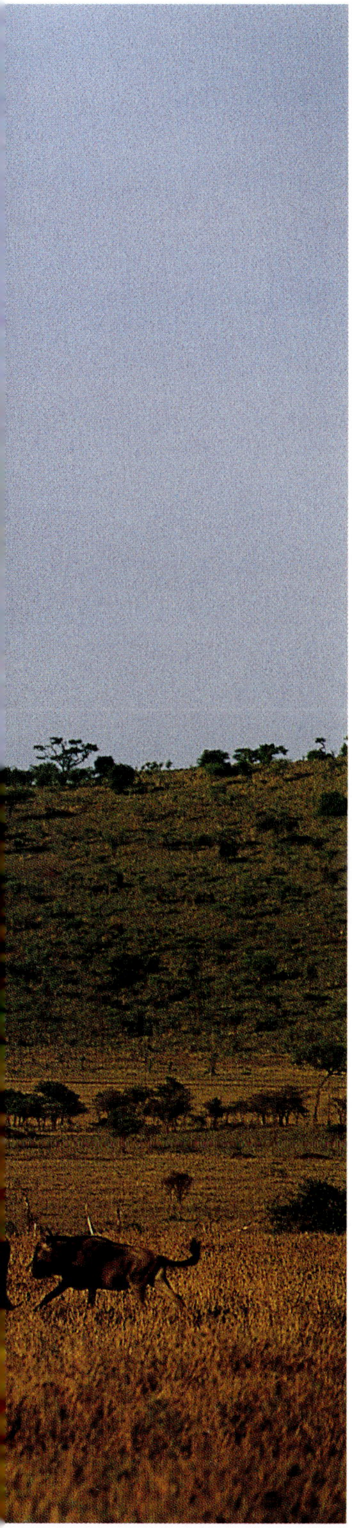

Crater rhinos

Ngorongoro Crater is a major tourist attraction (see Chapter 3); the gate fees alone bring in millions of dollars to the Ngorongoro Conservation Authority. Yet no amount in tourist dollars can ensure the long-term survival of one of the region's star attractions – the black rhino – in the crater landscape. More than 100 black rhinos roamed Ngorongoro Crater in 1965. In the intervening years, rhino numbers have plummeted, mostly as a direct result of poaching. In 1995, only 15 rhino remained. Then in 2001, four rhinos, out of a population of just 17 animals, died after being infected with a debilitating, tick-borne blood parasite, babesiosis. The outbreak of the disease occurred during a period of extreme drought, when a large number of other species, especially buffalo, had died in the crater. The parasite was eventually brought under control and the rhino population grew steadily; since January 2005, the birth of five calves has increased the population to 24 individuals.

For a small rhino population, such as that in the crater, local extinction by some unforeseen event, such as the devastating disease that struck in 2001, is not the only risk: there is also the danger of undermining the genetic diversity of the herd through inbreeding. Even though the population has reached its highest level in 30 years, the effective breeding population numbers only 13 animals.

Much of the crater's capacity to support rhinos is determined by the suitability of the habitat. Over the years, herbivore pressure, driven largely by elephants, together with an influx of saline water from the highly saline Lake Makat on the crater floor, has taken its toll on woody vegetation. The loss of wooded habitat, combined with the increased food requirements of elephant and buffalo, suggests that the crater can support, at most, only half of the former black rhino population.

The obstacles to survival are many: habitat shrinkage, loss of genetic diversity, drought, disease, tourist disturbance, calf predation by hyaenas, and the omnipresent threat of poaching all cast a shadow over the long-term future of the crater's rhinos.

Wildebeest

In a time before unfettered human fecundity dominated the environment, many species of animal kept a balance between their food supply and their appetite by migrating (see Chapter 4). Even elephant migrated on a scale to match their bulk, covering enormous distances. Wildebeest first appeared on the Serengeti plains around 1.5 million years ago and, as grazers, have evolved to live at high densities, exploiting an unstable food supply. In the 1880s, a cattle-borne disease known as rinderpest decimated the wildebeest population. A cattle vaccination programme was implemented in the 1950s and 1960s, and wildebeest numbers recovered; today's wildebeest population is thought to number some 1.6 million individuals.

Left: Tourists pay handsomely to take a balloon ride over the Serengeti plains. The migratory wildebeest often run when disturbed by low-flying, noisy balloons.

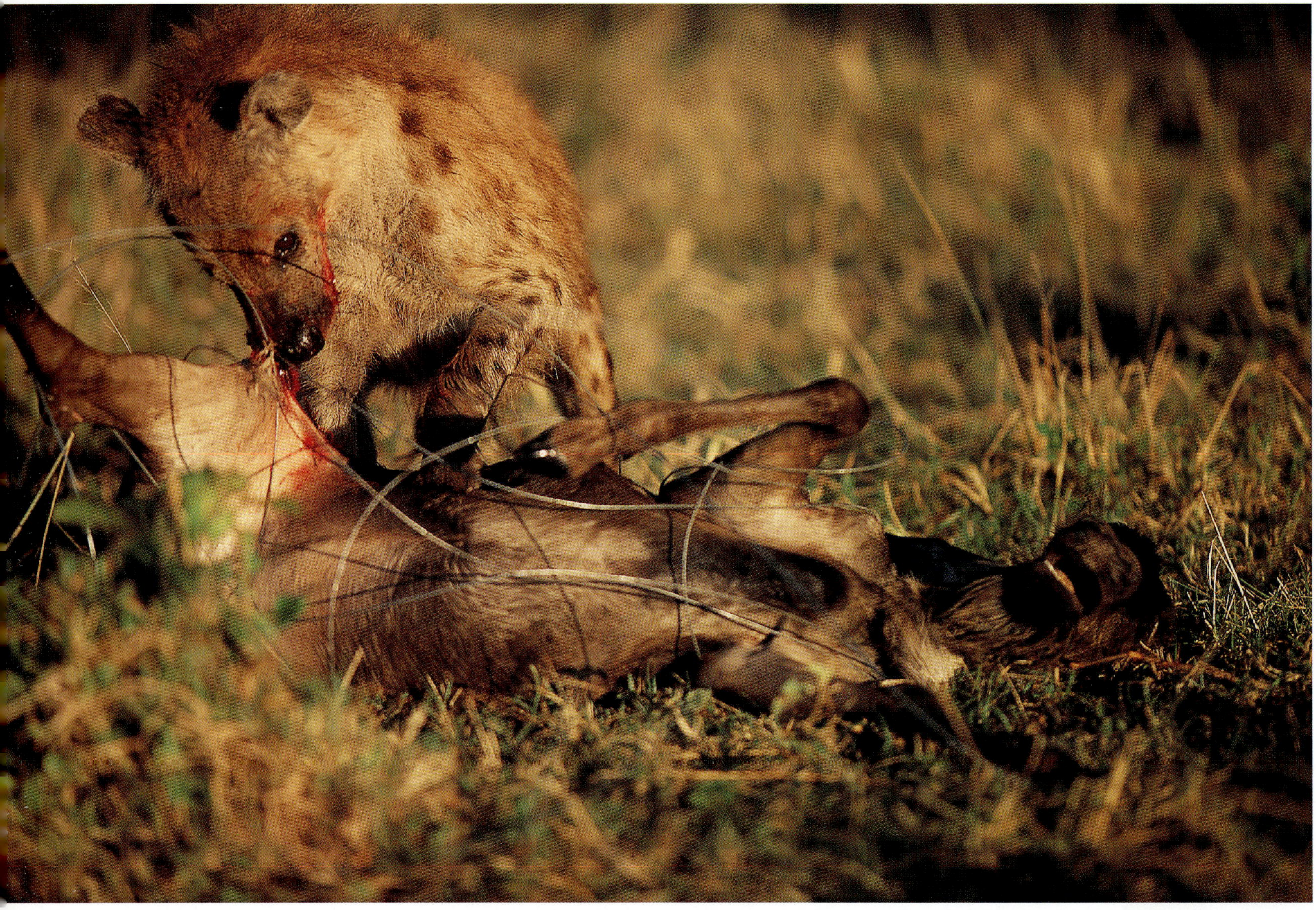

Poachers lay down snares in strategic places for ungulate meat. This indiscriminate and cruel method often snares carnivores; some animals are lucky to escape with a missing limb. Here a spotted hyaena has found an easy meal.

While most of the great land migrations of the world have been curtailed, the wildebeest migration still persists. Today, Serengeti National Park is far better protected from poachers than it was in the 1970s and 1980s, when the black rhinos were nearly eliminated and Serengeti's elephant population was reduced from 3 000 to 300. The migrant wildebeest, however, do not stay within protected areas; their migration takes them into areas of human settlements, especially in the west. Well over a million people are crowded between Lake Victoria and the western border of Serengeti National Park, bringing with them the accompanying, insistent demand for meat and land. The migrants are killed for the pot and for the market. Encroachment on the park by people is a serious problem. The wildebeest is a keystone species – one upon which every other species in the ecosystem depends. So, any threat to the animal is also a threat to many other species.

Mountain gorillas

The predicament of the mountain gorillas of the Virungas (see Chapter 5) is a direct result of the human urge to explore, to exploit, to 'manage', and, disgracefully, to hunt for fun or for profit; human interference has compromised the natural dignity of our ancestral cousins. At one time it seemed inevitable that mountain gorillas would be poached to extinction in the twentieth century. Initially European colonists hunted them for pleasure, killing indiscriminately; then the African population expanded and encroached on the gorillas' habitat; civil wars broke out, and the soldiers and refugees overran the Virungas. Despite all this, the mountain gorillas have survived into the twenty-first century and, almost miraculously, their numbers have increased. This has been achieved by the trick of turning them into a renewable resource for the tourism industry. Today mountain gorilla tourism brings in millions of dollars to the economies of Rwanda and Uganda, and there are plans to expand the industry further.

There are two adverse consequences. One is that over 70 per cent of Virunga gorillas are managed by humans: most are habituated groups visited by tourists; the remainder are habituated groups observed by researchers. Truly wild gorillas are becoming rarer, as the lure of the tourist dollar casts the net of human contact wider. The other consequence is that, because of their genetic closeness to humans, mountain gorillas are vulnerable to human diseases. They have little natural immunity against any human infection, and viruses or bacteria brought in by a tourist or a researcher could devastate a vulnerable gorilla group.

The stark reality is that the fate of the gorillas lies with us: we are both their nemesis and their salvation; we have total, absolute control over them.

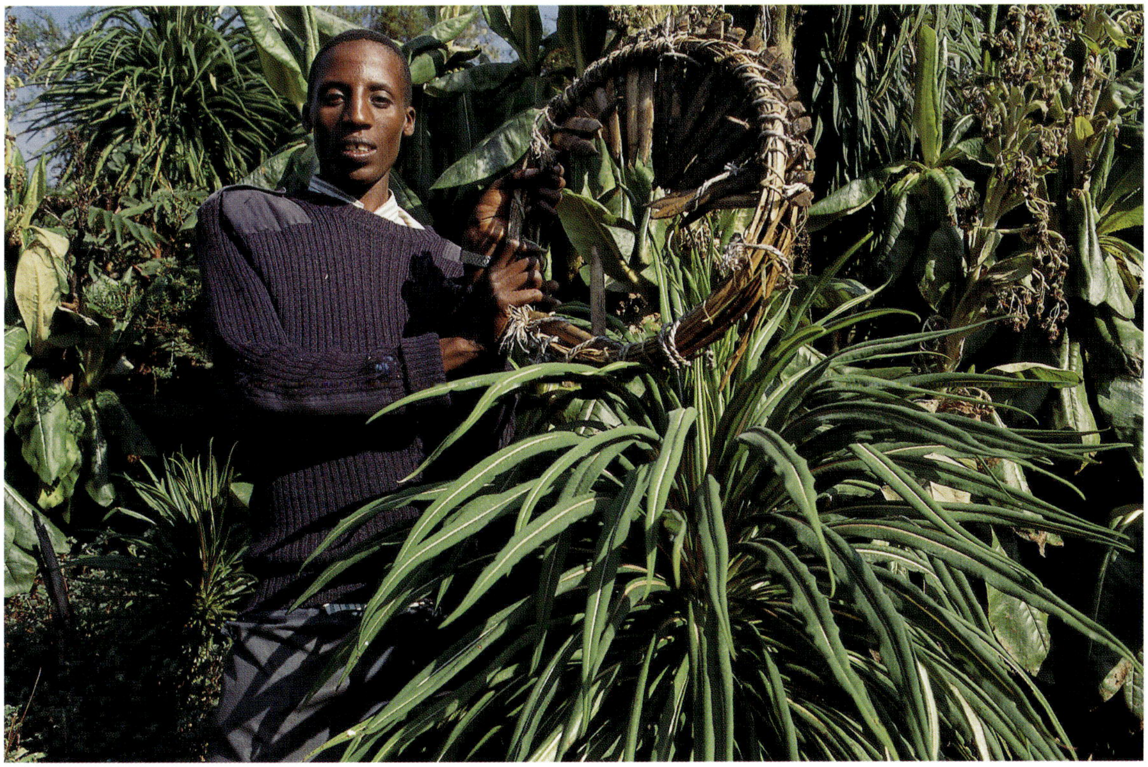

A ranger holds a snare found in the Virungas. Although the snares are intended for other prey, mountain gorillas do occasionally get caught in them, enduring unnecessary suffering.

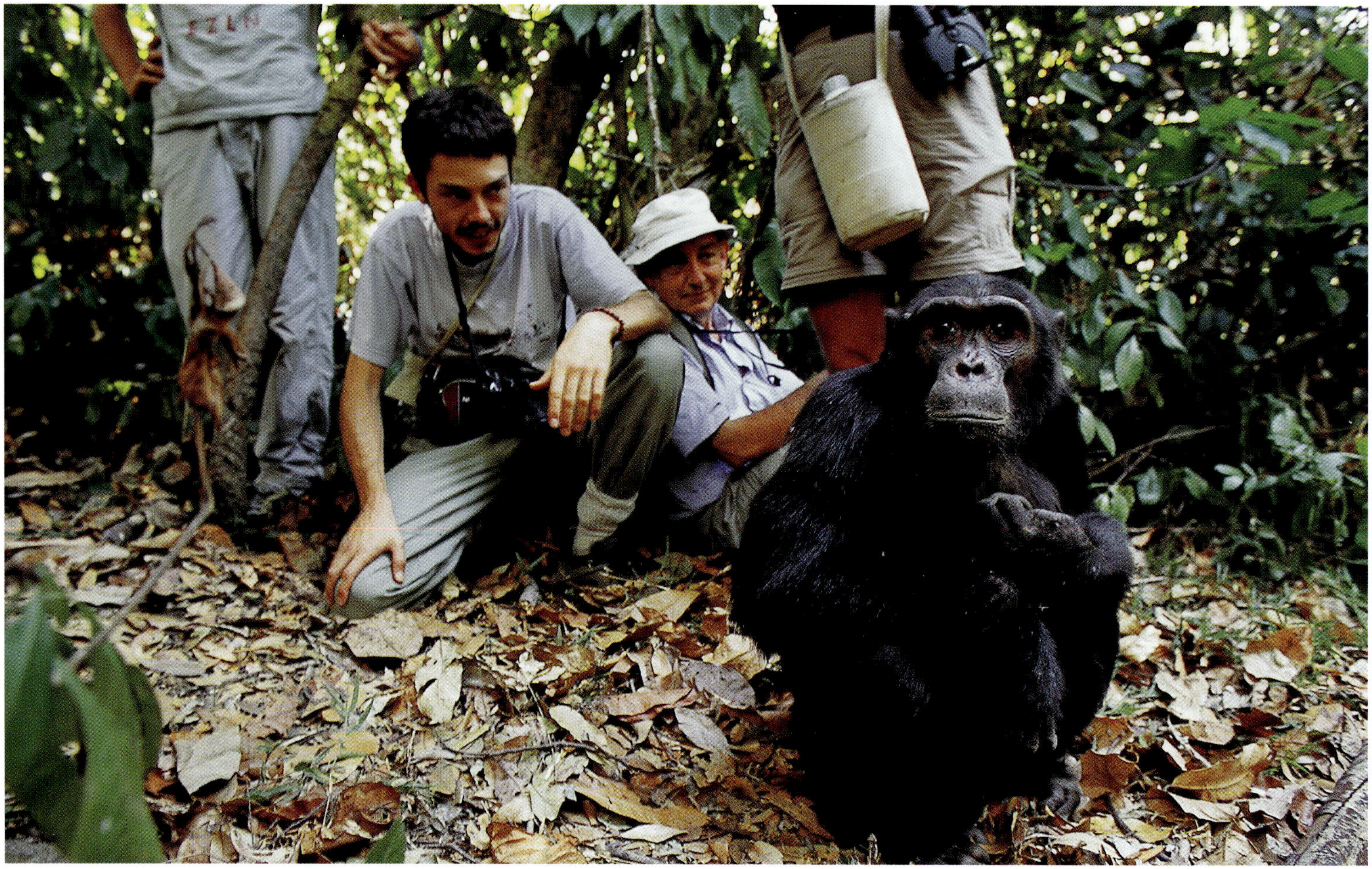

Adolescent Mahale chimpanzee, Cademus, is unperturbed as he stops to rest in front of an admiring tourist audience. Chimpanzee trekking is gaining popularity, increasing the risk of infecting chimpanzees with human diseases against which they have no immunity.

Chimpanzees

Gombe Stream National Park, known world-wide for its chimpanzees (see Chapter 6), is a sober lesson in the futility of expecting the power of publicity alone to conserve national parks. Championed by Jane Goodall, and the subject of several *National Geographic* articles and many more television documentaries, Gombe, as a chimpanzee habitat, is dying. The biggest short-term threat is the growing human pressure on the park.

Only 12.9 kilometres (eight miles) long and between 1.6 and 4.8 kilometres (one and three miles) wide, Gombe was in 1960 surrounded by large tracts of undisturbed forest stretching to the east and south, with strips of forest to the north linking the Gombe chimpanzees with those in nearby Burundi. Today, that habitat outside the park has all but vanished. Gombe has become a forest island, 35 square kilometres (13.7 square miles) in size, surrounded by farms, people and denuded hillsides. The loss of surrounding forest poses a great long-term threat to the Gombe chimpanzees. At one time they numbered around 150 chimpanzees in three communities. Today, that number has halved, and one community is on the verge of collapse. With falling numbers and a shrinking habitat, there is the imminent danger that the remaining Gombe chimpanzees will interbreed, lose genetic diversity and succumb to a fatal disease.

The hunter-gatherer way of life

It is ironic that while many in the West are advocating an environmentally benign lifestyle – practised so graciously by the traditional Hadzabe (see Chapter 8) – modern Tanzania is eradicating it in the name of development.

The Hadzabe survived the Bantu diaspora of the sixteenth century by retreating to Lake Eyasi Basin, a breeding ground for tsetse flies, which are the vectors for sleeping sickness, an infectious cattle disease. Because the Hadzabe are hunter-gatherers, rather than pastoralists, they had no cattle; they could tolerate the flies. They were able to satisfy all their needs directly from the land. They could light a fire, make poison, eat, sleep and be happy without recourse to material possessions. Their robust lifestyle was simple – fresh air, much exercise, the freedom to roam and freedom from the clock; leading a life free of stress, they had very little impact on their environment.

In the modern world, such an exemplary lifestyle is sadly forbidden. The land that they have lived on for centuries and looked after so well by living sustainably, maintaining the ecological balance, is now being redistributed. The Tanzanian government has recently promoted a land utilisation programme that favours pastoralism and farming, and is encouraging the eradication of tsetse flies so that pastoralists and others can move safely into the Hadzabe lands – but pastoralists and farmers have already invaded the area and altered the character of the environment. The Hadzabe have neither title deeds to their ancestral land, nor allies to fight their corner, since modern society places no value on the traditional Hadza lifestyle. The combined assaults on the land – vegetation clearance, charcoal burning, the grazing of cattle, goats and sheep, and the burgeoning demand for housing – are turning the region into a desert. The wildlife has retreated to protected areas on higher ground, where hunting is not allowed. The Hadzabe are losing their land, their livelihood and their lifestyle. The land itself is losing a friend, and the homogenised world is losing a precious heritage.

The land of Hadza hunter-gatherers is being overrun by domestic animals, with adverse effects on soil and vegetation.

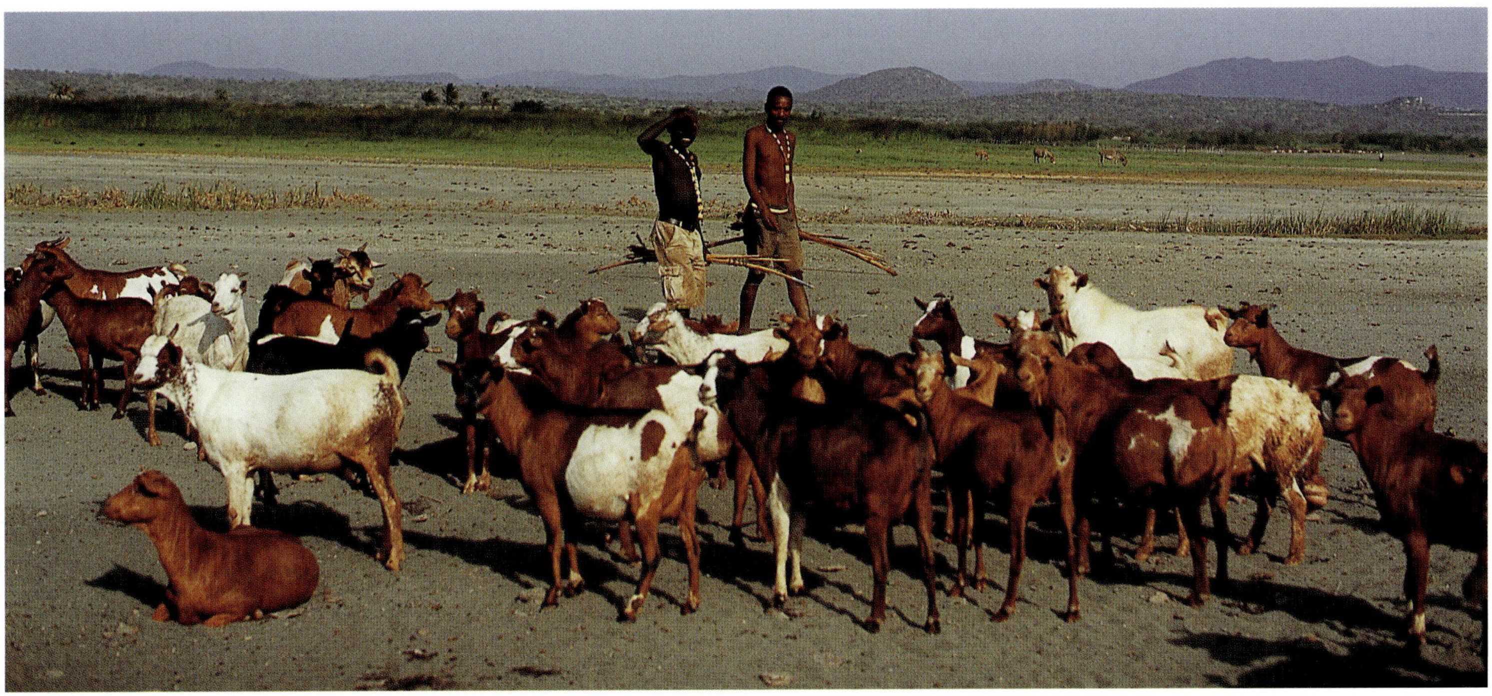

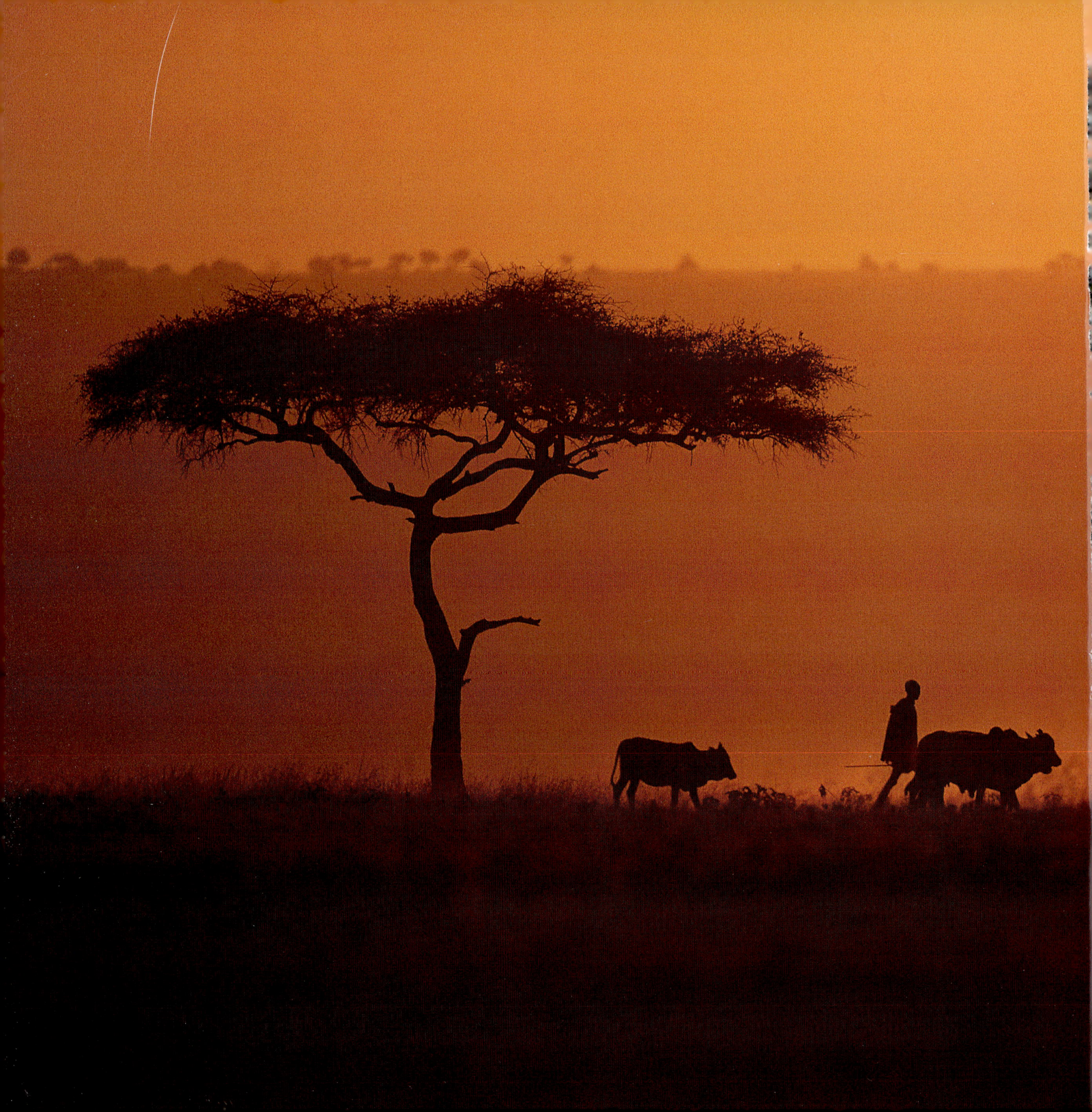